Economic and Environmental Benefits and Costs of Transgenic Crops: Ex-Ante Assessment

edited by

C.Ramasamy and K.N.Selvaraj

Tamil Nadu Agricultural University
Coimbatore, India

George W. Norton

Department of Agricultural and Applied Economics,
Virginia Polytechnic Institute and State University,
Blacksburg, Virginia Tech, USA

K.Vijayaraghavan

Sathguru Management Consultants Pvt Ltd
Hyderabad, India

"The debate about suitability of biotech agricultural products goes beyond issues of food safety. Access to biotech seeds by poor farmers is dilemma that will require interventions by governments and private sector. Seed companies can help improve access by offering preferential pricing for small quantities of biotech seeds to smaller farmers. Beyond that, public-private partnerships are needed to share R&D costs for "pro-poor" biotechnology."

Dr. Norman Borlaug, Nobel Laureate.

This publication has been compiled with support provided by Agricultural Biotechnology Support ProjectII ABSPII, a consortium led by Cornell University and supported by United States Agency for International Development (USAID).

ABSPII focuses on the safe and effective development and commercialization of bio-engineering crops as a complement to traditional and organic agricultural approaches in developing countries. The project helps boost food security, economic growth, nutrition and environmental quality in East and West Africa, Indonesia, India, Bangladesh and Philippines. Funded by the United States Agency for International Development (USAID) and led by Cornell University, ABSPII is consortium of public and private sector institutions.

The South Asia program is managed by Cornell University and Sathguru Management Consultants.

Economic and Environmental Benefits and Costs of Transgenic Crops: Ex-Ante Assessment

Printed and published by Tamil Nadu Agricultural University (TNAU).

Tamil Nadu Agricultural University acknowledges the support from United States Agency for International Development (USAID), Cornell University and Sathguru Management Consultants in bringing out this edited book.

Edited by:
C. Ramasamy and K.N.Selvaraj (Tamil Nadu Agricultural University, Coimbatore, India), George Norton (Virgina Tech, Blacksburg, US) and K.Vijayaraghavan, (Sathguru Management Consultants, Hyderabad, India)

ISBN: 978-81-904337-2-3

CONTENTS

Message by Mission Director, USAID – New Delhi

George Deikun

Agricultural Biotechnology Support ProjectII (ABSPII), a USAID funded consortium of Public and Private sector institutions that supports scientists, regulators, extension workers, farmers and the general public in developing countries to share knowledge on agricultural biotechnology. The consortium also focuses on the safe and effective development and commercialization of bioengineered crop as a complement to traditional and organic agricultural approaches. The project helps boost food security, economic growth, nutrition and environmental quality in East and West Africa, Indonesia, India, Bangladesh and the Philippines. The consortium develops innovative, pragmatic solutions, building on the success of ABSPI, which was led for over a decade by Michigan State University. The ABSPII project is led by the Cornell University.

ABSPII supported study titled "Economic and Environment Benefits and Costs Of Transgenic Crops : Ex-Ante Assessment", published by Tamil Nadu Agricultural University is a study of the impact of transgenic varieties of Rice, groundnut, potatoes and egg plant on the lives of the marginal farmers in South Asia.

There was a long felt need to address the issues pertaining to the needs of the farmers in the South Asia region. Conventional breeding methods in these areas were not able to address issues of increased exposure of women to pesticides, lesser yield and better crop returns for the farmers. Hence, under the guidance of Dr. George Norton from Virginia Tech University, researchers from Tamil Nadu Agricultural University and Bangladesh have conducted the Ex-ante studies on Socio economic impact of introducing the transgenic crops to this region on the farmers of India and Bangladesh. (Bt Eggplant, Drought and salt tolerant rice, virus resistant groundnut and sunflower, disease resistant potato). These study results clearly indicated that the introduction of transgenic crops improved overall returns to the farmers by reducing the use of pesticides and also increase in farm yield with better environment. It is a welcome sign that the marginal farmers of this region willing to adopt the state of the art, advanced biotech crops to extract maximum benefits from the modern technology.

The "Economic and Environmental Benefits and Costs of Transgenic Crops – Ex-Ante Assessment" documents that the introduction of the transgenic varieties will go a long way in helping to improve the lifestyle of the farmers in these regions. The introduction of the transgenic varieties has potential to

Accomplish higher farm yields and improved crop quality.

Contribute to increase in monetary compensation for the farmers as well as increased food security for the region

Decreased use of pesticide/fungicide for the better environment

Decreased exposure of women and children to chemicals and pesticides.

These results have explicitly exhibited the role of biotechnology in lifestyle changes and Socio-Economic improvement of the marginal farmers.

I hope this study will form the basis for making informed decisions in encouraging the biotechnology-derived foods for reduction of poverty and hunger and protection of environment.

Best wishes,
George Deikun

Message from ABSPII

Ronnie Coffman

Frank Shotkoski

ABSPII works to bring the benefits of agricultural biotechnology to Africa and Asia. Because we are a not-for-profit organization led by Cornell University and funded by the United States Agency for International Development (USAID), our consortium is able to focus on crops that are important to resource-poor farmers and consumers. As these crops may have limited markets or commercial value, they are often overlooked by multinational biotechnology companies. ABSPII works with local agricultural universities, private biotechnology companies, and seed companies to conduct research and development activities necessary to bring the highest quality products to farmers.

In 2002, stakeholder representatives from India and Bangladesh met to advise ABSPII on where to invest in biotechnology research for the region. Under this priority setting exercise, local stakeholders decided that chickpea, eggplant, groundnut, potato, sunflower and rice were the crops that were most appropriate candidates for ABSPII investment. The main criteria were that (1) the crop faced significant constraints to production that could be improved best with the help of biotechnology and (2) that addressing production constraints would have the greatest likelihood of improving quality of life for resource-constrained farmers and consumers.

ABSPII funded impact assessment studies to evaluate the market-level consequences of all our biotechnology products. Researchers at TNAU, BSMR Agricultural University (Bangladesh), the University of Hohenheim and Virginia Tech carried out the assessments. The results of these studies have helped us evaluate and refine our projects to ensure that we invest only in products with the greatest potential to help resource-poor farmers and consumers in our partnering countries.

While ABSPII is no longer involved in the development of genetically modified chickpea, groundnut or sunflower, we hope these studies will convince other organizations to continue this important product-development work. Research on eggplant, potato and rice are ongoing and we hope that the benefits of these improved crops will begin to reach farmers and consumers in India and Bangladesh over the next five to ten years.

Ronnie Coffman
Chair, International Programs
Cornell University, College of Agriculture and Life Sciences

Frank Shotkoski
Director, ABSPII

Author Profiles

George W. Norton is professor of Agricultural and Applied Economics at Virginia Tech where he has been on the faculty since 1980. He teaches courses on international agricultural development and trade, and conducts research on agricultural research evaluation, integrated pest management, and agricultural development issues. His current research projects focus on evaluating economic impacts of integrated pest management and of agricultural biotechnologies. He serves as impact assessment coordinator for the Agricultural Biotechnology Support ProjectII (ABSPII) and as chair of the Technical Committee of the Integrated Pest Management Collaborative Research Support Program (IPM CRSP). He was a visiting professor at Cornell University in 1987-88, and at the Economic Research Service, U.S. Department of Agriculture in 1999-2000. He held a courtesy appointment as a senior research fellow at ISNAR from 1987 to 1993. He received his BS from Cornell University and his MS and Ph.D. from the University of Minnesota. He has conducted research in more than 30 countries in Asia, Africa, Latin America, the Caribbean, and the Middle East.

C. Ramasamy is the Vice – Chancellor of Tamil Nadu Agricultural University, Coimbatore,India since December 2002. As an Agricultural Economist he has served TNAU in different positions for a period of 31 years. His major fields of specialization are Technology and Agricultural Development, international trade, and research investment economics. He has been the major advisor for 42 masters' scholars and 14 doctoral scholars. He has completed 50 research projects in collaboration with research institutes of national and international importance. He has published 70 research papers, six books, 51 chapters in books, two training manuals and 17 popular articles. He is a recipient of Ford Foundation Fellowship (twice), Winrock Fellowship, Rockefeller Fellowship, Best Researcher Award of TNAU (1995), D. K. Desai prize for the best research article (1997 and 2004) and Tamil Nadu Scientists Award (2001) from Tamil Nadu State Council for Science and Technology. He is a Fellow of the National Academy of Agricultural Sciences, New Delhi and a member in the Board for the Indo-US Knowledge Initiative for Agriculture. He is a member in the Project

Management Committee, National Agriculture Innovation Project, Indian Council for Agricultural Research, New Delhi and Nano-Biotechnology Committee, Ministry of Science and Technology, Government of India.

K.N.Selvaraj, Professor of Agricultural Economics was raised in a small village of Nilgiris District of Tamil Nadu, India and joined the University as Assistant Professor in the year 1989. He has obtained his Ph.D. in Agricultural Economics from Tamil Nadu Agricultural University in 1996. Since then he has been teaching graduate and post graduate students and guided around 20 students both in the capacity as Chairman and Members of the Advisory Committees for their theses research. He has been undertaking many research projects funded by various International and National funding organizations like Rockefeller Foundation, USA, Agricultural Economics Research Institute, Netherlands, Indian Council of Agricultural Research and National Centre for Economics and Policy Research. He has been trained in Technological Impact Assessment and International Trade in Virginia Technology, USA, Indian Institute of Foreign Trade, New Delhi, International Rice Research Institute, Philippines and other Institutes availing FAO fellowship and grant. He has participated in number of National and International meetings and visited various countries namely USA, Thailand, Mexico, Japan, France, Australia, Sri Lanka, Bangladesh, Italy, China, Philippines and South Africa. He has been involved in preparation of Policy Documents for State Government and served at various capacities in academic bodies.

Matin Qaim, a citizen of Germany, holds an MSc in Agricultural Sciences from the University of Kiel and a PhD in Agricultural Economics from the University of Bonn. From 2001 to 2003, he was a Post-Doc Visiting Fellow at the University of California at Berkeley (USA), before he became a Senior Researcher at the Center for Development Research in Bonn. Between 2004 and 2007, he was a Professor of Agricultural and Development Economics at the University of Hohenheim in Stuttgart. In 2007, he became Professor of International Food Economics and Rural Development at the Georg-August-University of Goettingen. Qaim has extensive research experience related to the economics of agricultural technologies in developing countries. In particular, he has implemented and coordinated numerous studies on the adoption and impacts of biotechnology in the small farm sector in countries of Asia, Africa, and Latin America. Moreover, he has carried out research on market implications and nutritional effects of new crop technologies and other policy interventions. Qaim has published widely in international scientific journals and books. His research has also been awarded with several academic prizes.

S.M. Fakhrul Islam is currently Associate Professor and Head in the Department of Agricultural Economics, Bangabandhu Sheikh Mujibur Rahman Agricultural University, Gazipur, Bangladesh. Dr. Fakhrul received B.S. and M.S. degree in agricultural economics from Bangladesh Agricultural University in 1985 and 1987, respectively. He received Ph.D degree in agricultural economics in 1995 from University of Philippines at Los Banos and completed post Ph.D in 1999 from Aristotle University of Greece. He also served in various national and international organizations like Planning Commission, Ministry of Livestock, Bangladesh Agricultural Research Institute, IRRI, ILRI, FAO, EC and DANIDA. Dr. Fakhrul is author of two books and published good number of articles in various national and international journals.Dr. Fakhrul received "Chancellors Award" in 1987 from president of Bangladesh for academic excellence. He also received National Science and Technology Fellowship, USAID scholarship for Ph. D studies, IRRI scholarship Ph. D dissertation research and Greek Government scholarship for Post Ph.D studies in natural resource and environmental economics. His research interest is impact assessment, project monitoring and evaluation, natural resource and environmental economics, livestock economics, agribusiness and international trade.

Vijesh V. Krishna is a research fellow at the University of Hohenheim (Stuttgart, Germany), from where he has secured his PhD in Agricultural Economics. He studied Agricultural Sciences and specialized in Agricultural and Resource Economics in the Kerala Agricultural University (Thrissur, India), the University of Agricultural Sciences

(Bangalore, India), and the University of Cambridge (UK). His doctoral research was an ex- ante evaluation of adoption, impacts and consumer acceptance of GM eggplant in India, under the supervision of Prof. Dr. Matin Qaim. His works are published both in national and international research journals. Apart from the research on economics of technology adoption in developing country agriculture, he has worked on biodiversity and traditional knowledge related issues in India.

Chapter **1**

Introduction and Methods

G.W. Norton, C. Ramasamy and K.N. Selvaraj

Introduction

This book presents the results of a series of studies to assess potential economic impacts of transgenic insect resistant (Bt) eggplant, drought and salt-tolerant (DST) rice, tobacco streak virus resistant groundnut and sunflower in India, and insect resistant (Bt) eggplant, late blight resistant (LBR) potato, pod borer resistant chickpea, and DST rice in Bangladesh. Research and development activities have been undertaken for the past four years under the auspices of the Agricultural Biotechnology Support ProgramII (ABSPII) for the purpose of commercializing products that solve major pest and other problems in the target commodities and countries[1]. This book summarizes the projected level and distribution of costs and benefits associated with those activities and products in South Asia.

Economic impacts of biotechnologies may be felt by producers, household members, consumers, and seed companies. Transgenic crops may differ from conventional crops in terms of costs, yields, potential impacts on health and the environment, and production and income risk. These effects may be felt unevenly by region, farm size, income level, gender, and consumers versus producers.

Modern biotechnology is a relatively young field, but research results have outpaced the ability of social scientists to adequately evaluate and explain benefits and costs to an often-wary public. Without informed public debate, potentially useful technologies are lumped together with potentially harmful ones, and approvals for important technologies may be delayed. Informing that debate requires economic analysis of the level and distribution of benefits and costs of transgenic crops, and a concerted effort to provide this information to the public.

---

[1] ABSPII is a USAID-funded project managed by Cornell University. Substantial support has also been provided by the Sathguru Management Consultants.

Often, economic assessments of improved technologies are conducted after the technologies have been released and the resulting products adopted. Such assessments provide an accounting of the benefits from the research investments. However, equally important can be "ex-ante" assessments of the potential benefits of technologies not yet released and adopted. These assessments can provide information to help guide resource allocation decisions and justify continued funding for on going research programs. In the case of biotechnologies, they can also indicate where emphasis might be placed in the regulatory progress to maximize return on investment.

This book presents the results of a set of ex-ante impact studies, conducted with support from ABSPII, for the purpose of providing information to project leaders, funding agencies, policy makers, and the public officials that will help them make resource allocation decisions and rational choices about supporting various transgenic crops in South Asia[2]. After discussing the basic methods employed in the studies, Chapters 2 to 9 summarize the projected direct economic impacts of biotechnologies for specific commodities and countries. Chapter 10 summarizes and concludes the book.

Methods

A basic economic impact analysis was conducted for each technology/crop to project benefits of the technology using a consistent framework that included: (a) review of existing data on crop losses and cropping practices to manage the targeted problems in major locations in India and Bangladesh where the crops are grown, (b) obtain crop budgets, experimental data, and opinions of scientific experts to budget out expected per hectare cost and yield changes with and without the transgenic technologies, (c) gather information on time and cost required to complete the research and meet regulatory hurdles, (d) gather data on outputs, output prices, and international trade of each commodity and assess the nature of the market (e.g., closed, small-open, or large-open economy) (e) assess rate and timing of adoption of transgenic varieties, and (f) conduct ex-ante economic surplus analysis. A basic environmental assessment included obtaining estimates of changes in pesticide use for each commodity where it was expected to change with the technology.

---

[2]A companion book is being prepared for ABSPII impact assessment studies in South East Asia.

Data Review on crop losses and cropping practices – Existing published and other data on crop losses and cropping practices associated with the problem targeted by the transformation were reviewed for the relevant crops and countries. Data were obtained for major locations where the crops are grown.

Budgeting – Data were gathered on yields and input costs from field trials or experimental data for the transgenic and alternative technologies, and opinions of biological scientists, other industry experts, and farmers were solicited (See Appendix 1 for sample surveys). Published data (for the last four years) on prices, production, and trade nationally and regionally for the target crops were collected, and decisions were made on the nature of the markets. Budgets were constructed for each target crop and technology for the relevant regions.

Time to and rate of adoption – The percentage and timing of adoption of the transgenic varieties by region were projected based on information on agro-ecological, socio-economic, variety considerations, and other factors, including information on where the targeted problem is most severe. Adoption of previous technologies were considered, as well as expert opinion on factors such as the research lag, lags due to the regulatory process, and projections on how the seeds will be commercialized to arrive at estimates of the likely timing and rate adoption by farmers in different regions.

Economic Surplus Analysis – Estimates of price elasticities of supply and demand were obtained for each target crop based on published estimates or on economic theory Scientists were asked to assess the probability of achieving technical success with the research. The budget information, secondary data, and adoption information were combined in an economic surplus model to assess the total economic benefits and their distribution by region within each country and to consumers, producers, and seed sector. The costs of the research and product development (including meeting regulatory hurdles) were included along with the benefits in a benefit cost analysis of the public investment.

As described in Norton. et al., (2005), when widespread adoption of a new technology occurs across large areas, changes in crop prices, cropping patterns, producer profits, and societal welfare can occur. These changes arise because costs differ and because supplies may increase, affecting prices for producers and consumers. These changes are illustrated in Figure 1. In this figure, S0 represents the supply curve before adoption of a new GMO, and D represents the demand curve. The initial price and quantity are P_0 and Q0. Suppose the new technology leads to savings of R in the average and marginal cost of production, reflected in a shift down in the supply curve to S1. This shift leads to an increase in production and consumption of Q1 (by $\Delta Q = Q1 - Q0$) and the market price falls to P1 (by $\Delta P = P0 - P1$). Consumers are better off because they can consume more of the commodity at a lower price. Consumers benefit from the lower price by an amount equal to their cost saving on the original quantity ($Q0 \times \Delta P$) plus their net benefits from the increment to consumption. Total consumer benefits are represented by the area P_0abP1.

Although they may receive a lower price per unit, producers are better off too, because their costs have fallen by R per unit, an amount greater than the fall in price. Producers gain the increase in profits on the original quantity ($Q0 \times (R-\Delta P)$) plus the profits earned on the additional output, for a total producer gain of P1bcd. Total benefits are obtained as the sum of producer and consumer benefits.

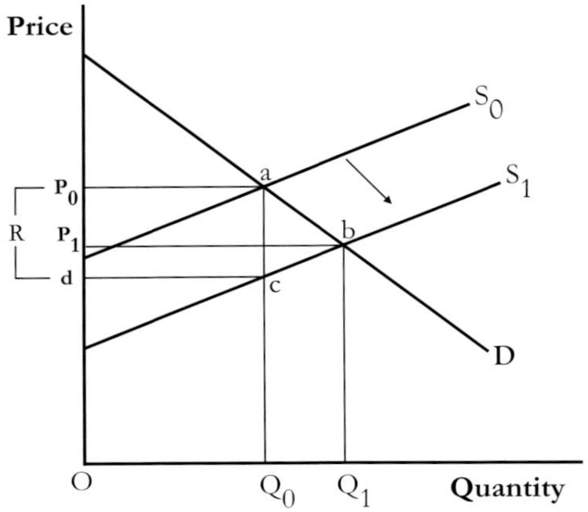

Figure 1.
GMO Benefits Measured as Changes in Economic Surplus

The distribution of benefits between producers and consumers depends on the size of the fall in price (P) relative to the fall in costs (R) and on the nature of the supply shift. For example, if a commodity is traded and production in the area producing the commodity has little effect on price, most of the benefits would accrue to producers. If the supply curve shifts in more of a pivotal fashion as opposed to a parallel fashion as illustrated in Figure 1, the benefits to producers would be reduced. Formulas for calculating consumer and producer gains for a variety of market situations are found in Alston. et al., (1995). For example the formula to measure the total economic benefits to producers and consumers in Figure 1, which assumes no trade, is KP0Q0 (1 + 0.5Zn), where: K = the proportionate cost change, P0 = initial price, Q0 = initial quantity, Z = Ke/(e +n), e = the supply elasticity, and n = the demand elasticity. Other formulas would be appropriate for other market situations. The two major market situations addressed in the chapters in this book are first a closed economy (no trade) and second an open economy in which the country cannot affect the world price of the product even though it trades it.

Once changes in economic surplus are calculated or projected over time, benefit/cost analysis can be completed in which net present values; internal rates of return, or benefit cost ratios are calculated. The benefits are the change in total economic surplus calculated for each year, and the costs are the public expenditures on the research and regulatory process. The primary purpose of the benefit/cost analysis is to take into account the fact benefits and costs need to be discounted, as the sooner they occur the more they are worth. The net present value (NPV) of discounted benefits and costs can be calculated as follows:

$$NPV = \sum_{t=1}^{T} \frac{R_t - C_t}{(1+i)^t}$$

Where: Rt = the return in year t = change in economic surplus

Ct = the cost in year t (the IPM program costs)

i = the discount rate (in most cases assumed to be 5% in this study)

Economic surplus estimates and net present values for this study were calculated using Excel spreadsheets. The underlying spreadsheets for each product are available from the authors of the individual chapters. In summary, a standardized set of methods were used to evaluate the impacts of transgenic crops in India and Bangladesh.

Reference

Alston, Norton G., K. Moore, D. Quishpe, V. Barrera, T. Debass, S. Moyo and D. Taylor. "Evaluating Socio-Economic Impacts of IPM." Chapter 12, pp 225-244, in G.W. Norton, E.A. Heinrichs, G.C. Luther, and M.E. Irwin (eds.) *Globalizing IPM: A Participatory Research Process Blackwell Publishing.* Ames, Iowa, 2005.

Chapter **2**

Drought and Salinity Tolerant Rice

C. Ramasamy, K.N. Selvaraj and G.W. Norton

Introduction

Rice consumption will increase in the future in India due to both population growth and increases in per capita incomes. Rice production may need to increase as much as 70 percent over the next two decades to meet this growing demand. This production increase must be achieved with less land and water. Drought and salinity are abiotic stresses that seriously hinder rice production in India, especially in areas where populations are most vulnerable. Drought is especially severe in rain-fed areas, raising the risk for those already close to the margin. Similarly, salinity reduces both yield and the area cultivated. India, with the help of the public and private sectors, has been developing drought and salt tolerant rice with the potential to increase production and reduce its variability. This genetically modified rice is still several years away from commercialization, but the purpose of this study is to project the potential economic value that such rice would have for India.

Rice Production in India

Among the food grains, rice occupies a dominant position in India. The country has the largest acreage under rice (43.70 million hectares during 2006-07) in the world, and its 90 million tonnes (91.05 million tonnes during 2006-07) ranks second only to China. The post independence era has witnessed spectacular progress in rice production and productivity. From 1950-51 to 2006-07, the area increased by one and half times (30.81 to 43.70 million hectares), yield by three times (668 to 2084 kg per hectare) and production by four and half times (20.58 to 91.05 million tonnes).

At the current rate of population and income growth, rice production must increase to about 125 million tonnes by 2020 (Mishra, 2005). The country must produce an additional 2.33 million tonnes per annum to achieve the target. Achieving this target is a major challenge, as this increase must be attained with shrinking land and water resources, scarce and costly labour and other inputs and a deteriorating environment. Further, due to technological stagnation in recent years and environmental constraints such as water logging, drought, and soil salinity, rice productivity has stagnated even in the highly favorable irrigated areas, while in some pockets yield has declined, causing great concern for food security (Table 1).

TABLE 1.
Rice Productivity Growth during the 1990s in India

Productivity Growth	No. of States
Declining Productivity	5
Declining Marginal Productivity	13
Increasing Marginal Productivity	2

Source: Authors' estimates using secondary data published in Agricultural Statistics at a Glance, Agricultural Statistics Division, Directorate of Economics and Statistics, Department of Agriculture and Cooperation, Ministry of Agriculture, Government of India, New Delhi.

Rainfed Rice Production

Technological change and infrastructure, especially irrigation, are the most important factors that have contributed to the rapid growth in rice production over the past 35 years (Hazell and Ramasamy, 1991, Pingali et al., 1997; Pingali and Hossain, 1999; Batia 1999). Area under rice in 2006-07 was 43.70 million hectares, of which West Bengal accounted for 13. 24 percent followed by Uttar Pradesh (12.78 percent), Orissa (10.26 percent) and Andhra Pradesh (9.12 percent). Rice is the staple food of 65 percent of the population in India. It constitutes about 43 percent of the total food grain production and 55 percent of total cereal production. India became self sufficient in rice in 1977, which was achieved through increased area under cultivation, increased cropping intensity, and wide-scale adoption of modern varieties (MVs). As a result, an average annual increase of over 2 percent in rice yield has been attained. Rice production has exceeded 50 million tonnes annually since the 1970s, and total rice production in 2006-07 was 91.05 million tonnes with an average yield of over 2 tonnes per hectare. Rice yield increased from 1341.53 kg per ha in the 1970s to 2038.86 kg per ha in the 1990s (Table 2 and Figure 1). However, rice yields remain below 2 tonnes per ha in many of the rice growing states despite a sizeable area under high yielding varieties in those states.

Rice environments in India are extremely diverse (Table 3 and Figure 2). Of the over 40 million ha of harvested rice area, about 33 percent are rain fed lowland, 45 percent irrigated, 15 percent rain fed upland, and 7 percent flood-prone. Since the major portion (55 percent) of the area under rice in India is rain fed, production is strongly tied to the distribution of rainfall. In some of the states, erratic rainfall leads to drought during the vegetative period, but later on the crop may be damaged by submergence due to high rainfall. Irrigated ecology accounts for the highest production and productivity closely followed by rain fed shallow lowlands. The irrigated north and south zones together account for 39 percent of the area under rice in the country, slightly less than the eastern zone, but in terms of production these two zones contribute over 50 percent, which is almost one and a half times more than that of eastern India due to a distinct yield edge. The rain fed eastern zone (West Bengal, Orissa, Bihar, Assam, and Manipur) accounts for just over 40 percent of the total rice area and 35 percent total rice production in the country. Rain fed upland, just one half of the rain fed lowland area, produces less than one fifth of it (Table 4 and Figure 3). Rice productivity in most of the states in Eastern Zone is less than 2 tonnes per ha except in West Bengal. The relationship between yield and the major growth factors reveals that area under High Yielding Varieties (HYVs), with a high level of fertilizer consumption and irrigated rice area are the key factors responsible for high yield.

TABLE 2.
Growth in Rice Area, Production, and Yield in India (Percent)

Zone / State	Area (M.Ha)			Production (MT)			Yield (kgs/ha)		
	1970s	1980s	1990s	1970s	1980s	1990s	1970s	1980s	1990s
South Zone									
Tamil Nadu	2.61	2.14	2.24	5.45	5.73	8.1	2065	2712	3562
	(0.28)	(-1.85)	(0.39)	(-14.02)	(7.36)	(1.66)	(-14.28)	(9.38)	(0.73)
Andhra Pradesh	3.47	3.81	3.85	6.55	9.32	10.82	1898	2404	2793
	(2.37)	(1.42)	(0.67)	(13.86)	(6.68)	(2.08)	(0.30)	(5.49)	(1.40)
Karnataka	1.13	1.14	1.34	2.48	2.49	3.58	2215	2162	2625
	(-0.03)	(0.81)	(2.11)	(0.43)	(4.13)	(3.11)	(0.52)	(3.29)	(1.38)
North Zone									
Punjab	0.69	1.59	2.31	2.01	5.63	8.53	2795	3457	3621
	(12.47)	(5.69)	(2.74)	(17.26)	(11.02)	(2.53)	(4.24)	(5.06)	(-0.21)
Haryana	0.34	0.56	0.87	0.79	1.6	2.48	2285	2829	2805
	(6.50)	(3.36)	(6.09)	(10.59)	(6.97)	(3.11)	(3.51)	(3.51)	(-2.28)
Uttar Pradesh	4.74	5.31	5.68	4.78	8.4	12.1	858.40	1594	2100
	(1.02)	(0.05)	(0.82)	(0.12)	(10.19)	(2.15)	(-0.85)	(10.14)	(1.31)
East Zone									
West Bengal	5.18	5.3	5.81	7.54	9.4	13.78	1449	1751	2230
	(-0.26)	(1.12)	(0.13)	(-0.52)	(10.62)	(1.94)	(0.27)	(9.41)	(1.47)
West Zone									
Madhya Pradesh	4.58	4.94	4.94	3.9	4.77	5.7	717.40	866.90	1021.36
	(-0.92)	(0.43)	(-4.62)	(-3.35)	(5.30)	(-8.29)	(-4.23)	(4.84)	(-3.84)
Maharashtra	1.41	1.5	1.52	2.1	2.48	2.6	1485	1669	1696
	(1.66)	(-0.19)	(-0.61)	(3.75)	(3.08)	(-0.27)	(2.05)	(3.30)	(0.45)
India	38.91	40.40	43.42	52.19	58.42	88.53	1341.53	1446.22	2038.86
	(0.82)	(0.14	(0.66)	(3.51)	(-2.04)	(1.56)	(0.99)	(-1.37)	(0.88)

(Figures in parenthesis are estimated exponential growth rates)

Source: Authors' estimates using published data from Agricultural Statistics at Glance, Agricultural Statistics Division, Directorate of Economics and Statistics, Department of Agriculture and Cooperation, Ministry of Agriculture, Government of India, New Delhi.

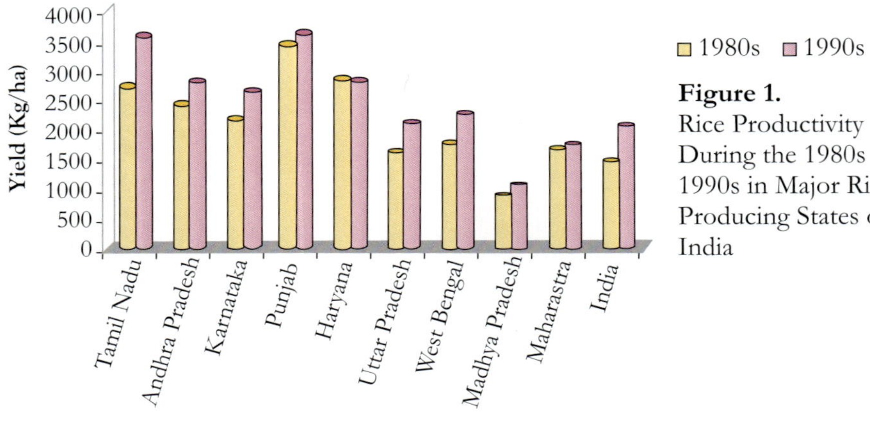

Figure 1.
Rice Productivity During the 1980s and 1990s in Major Rice Producing States of India

TABLE 3.
Percent of Rice Area under Rainfed Production

Year	1970s*	1980s*	1990s*
South Zone			
Tamil Nadu	8.28	7.88	8.65
Andhra Pradesh	5.63	6.43	7.38
Karnataka	36.85	38.09	33.68
North Zone			
Punjab	5.71	1.45	1.19
Haryana	9.10	2.12	1.49
Uttar Pradesh	79.32	67.07	45.80
East Zone			
West Bengal	73.61	75.12	75.21
Orissa	73.57	67.77	63.24
West Zone			
Madhya Pradesh	85.18	75.74	77.23
Maharashtra	75.65	75.60	73.65
India	60.66	56.75	51.21

*Averages for respective periods

Source: Authors' estimates using data from various published sources such as Agricultural Statistics at a Glance and Indian Agriculture in Brief, Agricultural Statistics Division, Directorate of Economics and Statistics, Department of Agriculture and Cooperation, Ministry of Agriculture, Government of India, New Delhi.

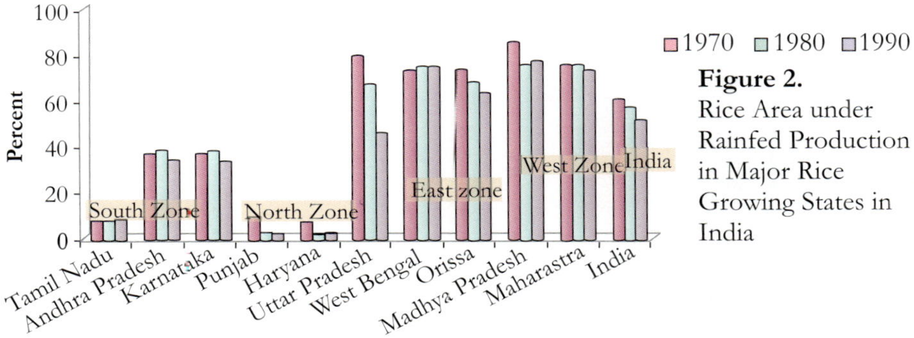

Figure 2.
Rice Area under Rainfed Production in Major Rice Growing States in India

TABLE 4.
Million tonnes of Rice and Percent Share of Rice Production by Zone in India

Zone / State	1970s	1980s	1990s	Overall
South Zone				
Tamil Nadu	5.18	5.11	6.76	5.72
	(11.64)	(17.52)	(8.39)	(9.38)
Andhra Pradesh	5.58	8.01	9.88	7.89
	(12.46)	(14.75)	(12.27)	(12.35)
Karnataka	2.05	2.24	3.20	2.52
	(4.59)	(6.84)	(3.97)	(5.1)
North Zone				
Punjab	1.67	4.92	7.59	4.82
	(3.67)	(6.9)	(9.4)	(5.97)
Haryana	0.70	1.38	2.21	1.46
	(1.53)	(4.04)	(2.74)	(2.77)
Uttar Pradesh	4.04	7.18	10.95	7.51
	(8.96)	(11.89)	(13.57)	(10.77)
East Zone				
West Bengal	6.36	8.15	12.31	9.05
	(14.27)	(18.34)	(15.29)	(15.17)
West Zone				
Madhya Pradesh	3.30	4.28	5.22	4.30
	(7.35)	(5.74)	(6.53)	(6.12)
Maharashtra	1.74	2.18	2.38	2.11
	(3.84)	(6.42)	(2.97)	(4.36)
India	44.76	54.65	80.53	60.64
	(100)	(100)	(100)	(100)

(Figures in parentheses are percentages)

Source: Authors' Estimates from several published sources such as Agricultural Statistics at a Glance and Agricultural Brief, Agricultural Statistics Division, Directorate of Economics and Statistics, Department of Agriculture and Cooperation, Ministry of Agriculture, Government of India, New Delhi.

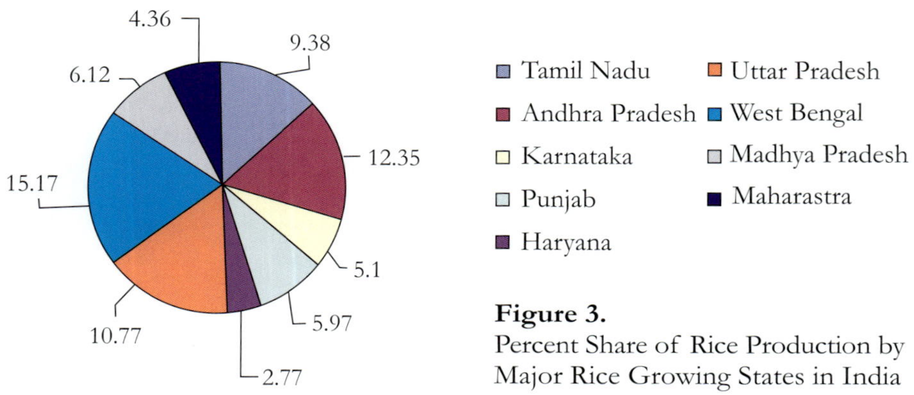

Figure 3.
Percent Share of Rice Production by Major Rice Growing States in India

Considerable yield variation among the rice production environments exists due to differential levels of adoption of new rice technologies, varying degrees of water control, imbalances in infrastructure development and a host of other factors. The yield difference between irrigated and non-irrigated areas indicates that rice yield under irrigation was more than 50 percent higher than in the non-irrigated areas (Table 5). Lower yield growth of rice in rain fed areas is mainly attributed to low spread of HYVs and purchased inputs such as fertilizer. Cultivation of low-yielding crop cultivars in marginal lands and absence of a major break through in the development of drought resistance is responsible for poor yields. Among the states in the southern zone, the yield gap in Tamil Nadu was 826 kg per hectare followed by Karnataka with 1794 kg per hectare. Even though the experimental yield potential of the states in the North Zone was higher than the states in the Southern zone, the yield gap was also higher. The maximum yield gap was noticed in Uttar Pradesh with 3728 kg per hectare, which accounts for 56.5 percent of the experimental yield. The yield gap in Haryana was 3322 kg per hectare, which represented 45 percent of the yield potential. Punjab, the leading state in agriculture in the country, was also experiencing a yield gap in rice cultivation. The average productivity of the states in the Eastern Zone was less than 2 tonnes per hectare except in West Bengal. The yield gap in the states of the Western Zone was more than 60 percent of the potential yield except in Maharashtra (Table 6).

TABLE 5.
Rice Productivity in Irrigated and Non-irrigated Areas in India (kg/ha)

Zone/State	Irrigated	Non-irrigated	Percent difference
South Zone			
Tamil Nadu	3263	1092	66.53
Andhra Pradesh	2978	722	75.75
Karnataka	2236	1370	38.73
North Zone			
Punjab	3545	–	–
Haryana	2652	–	–
Uttar Pradesh	2120	877	58.63

Source: Compiled from various sources of Agricultural Statistics at a Glance and Indian Agriculture in Brief, Agricultural Statistics Division, Directorate of Economics and Statistics, Department of Agriculture and Cooperation, Ministry of Agriculture, Government of India, New Delhi.

TABLE 6.
Yield Gap in Major Rice Growing States in India*

Zone/State	State Average (kg/ha)	Experimental Trial Average**	Yield Difference	Gap of State Average over Experimental Average (percent)
South Zone				
Tamil Nadu	4460	5286	826	15.6
Andhra Pradesh	3767	5882	2115	36.0
Karnataka	3456	5250	1794	34.2
North Zone				
Punjab	5042	6460	1418	22.0
Haryana	4074	7396	3322	44.9
Uttar Pradesh	2870	6598	3728	56.5
East Zone				
WestBengal	3147	5003	1856	37.1
West Zone				
Madhya Pradesh	1581	4710	3129	66.4
Maharashtra	2380	4501	2121	47.1
India	2759	5781	3022	52.3

* 1990-91 to 1997-98, ** Mean yield of best entry (irrigated medium at AICRIP test locations over 7 years period).
Source: Siddiq E.A., Rice: Yawing Productivity Gaps Survey of Indian Agriculture 2000, The Hindu, P.39

Spread of Modern Rice Varieties in the Rainfed Environment

Since the 1960s, the Indian agricultural research system has released many improved varieties (more than 600) some of which have been widely adopted by farmers (www.fao.org). The rice area which was planted to modern varieties during 1970s increased to more than 70 per cent in the 1990s. More than 80 per cent of the rice area was planted to modern varieties in the Punjab followed by Tamil Nadu (82 per cent), Andhra Pradesh (54 per cent) and Haryana (52 per cent) during the 1970s. Farmers' ability to invest in cost-intensive innovative technologies such as HYVs and modern inputs is limited in the rain fed environment. As a consequence, the range of technologies likely to be adopted by rain fed rice farmers is restricted. Rice is predominantly cultivated in rain fed condition in West Bengal, Madhya Pradesh and Maharashtra due low irrigation potential. Some of the local varieties are still popular there due to their tolerance and resistance to biotic and abiotic stresses.

Rainfall is the major source of water in the rain fed ecosystem and any uncertainty in rainfall distribution affects rice cultivation. Since farmers in these fragile environments experience frequent drought and rainfall failure, they mostly cultivate varieties that can assure at least a minimum yield during the extreme period. Studies have also shown that although more rice area is under canal irrigation in the rain fed ecosystems, the area under modern varieties is less because of the unreliability of canal irrigation. Even in the irrigated environment, adoption of modern varieties decreases with increases in variability of rainfall.

Rainfall Variability and Input risk

Much output variability is either due to weather or to insects and diseases. According to the report of the National Commission of Agriculture (Government of India, 1976), rainfall fluctuations could be responsible for 50 per cent of variability in yields. In the case of rice, the distribution of rainfall during the crop-growing season is found to be the most crucial weather parameter. The role of weather factors in crop growth often means that short duration varieties have lower climate induced variability than long duration varieties. The coefficients of variation of yearly rainfall in the various states of the country are presented in Table 7. Rainfall variability even in the irrigated

TABLE 7.
Rainfall Variability in Rice Production Zones (CV: Percent)

Zone/State	1970s	1980s	1990s*
South Zone			
Tamil Nadu	7.58	11.12	17.82
Andhra Pradesh	45.10	29.52	29.42
Karnataka	54.58	20.10	10.06
North Zone			
Punjab	52.95	31.06	23.46
Haryana	66.78	43.40	30.27
Uttar Pradesh	82.23	31.52	14.29
East Zone			
West Bengal	55.77	25.63	48.50
West Zone			
Madhya Pradesh	66.22	21.00	17.40
Maharashtra	66.68	35.64	20.47

* Averages of the respective period
Source: Authors' Estimates based on published data from Indian Ministry of Agriculture and Cooperation.

areas is as high as 40 percent to 60 percent during 1970s. Though the rainfall variability has not increased much over the years, it is affecting rice production in both rice production environments. With large-scale development of both surface and major irrigation systems, and implementation of watershed programs in many of the states, area under rain fed cultivation is declining in most of the states and area under irrigation is exhibiting positive growth in the rain fed ecosystems.

Drought and Salinity

The effect of drought is multidimensional. A literature review pertaining to the extent of loss in yield of rice due to drought was undertaken. Yield loss of rice in India due to drought ranged from 17 to 37 percent (Table 8). If rice production is the major source of income and employment, then a decline in rice production is not the only consequence of drought. It is estimated that the proportionate reduction in per capita consumption of the bottom 20 percent of the income distribution is 10 times that of the top 5 percent as income drops (Mellor, 1978). The most serious impact of drought is on the earnings of agricultural laborers, who make up about one-third of the rural population. When a crop is struck by drought, farmers may have no option but to cut it and sell it as feed for cattle. For agricultural laborers, this means not only work at a fraction of the normal wage rate, but also the disappearance of an entire chain of post harvest operations that would have given them a daily cash flow throughout the period. Migration is often common among the households in the rain fed region due to crop failure.

TABLE 8.
Rice Productivity Loss per Hectare Due to Drought

Region/State	Percent *
Dry season irrigated rice (West Bengal)	36.8
West Bengal	27.9
Tamil Nadu	29.0
Andhra Pradesh	23.0
Karnataka	18.0
Eastern India	16.9

* Based on earlier studies

In order to indicate the magnitude of loss in rice production due to the frequent occurrence of drought in India, an attempt was made to estimate losses in Tamil Nadu. A shortfall in rainfall occurred during the years 1974-75, 1980-81, 1986-87, 1988-89, 1990-91, 1995-96, 2002-03, and 2003-04. In other words, there were eight drought years in Tamil Nadu during the last 35 years (1970-71 to 2003-04). There were seven drought years in the rain-fed rice production environment over the last 35 years (Table 9 and Figure 4). During the drought period, the average rainfall was 694 mm in Tamil Nadu, while it was 965 mm in the normal period, a shortfall in rainfall of 39 percent as compared to a normal year. The shortfall in rainfall was 53 per cent in rain fed rice production during the drought period as compared to rainfall during a normal period (Table 10).

TABLE 9.
Number of Drought and Normal Years

S.No	Districts	Drought Year (>-20%)	Normal Year (<-19.9%)
1	RF	7	29
2	Tamil Nadu	8	28

RF – Rainfed Production Environment
Source: Authors' estimates based on Government data

Figure 4.
Rice Production Shortfall, Tamil Nadu

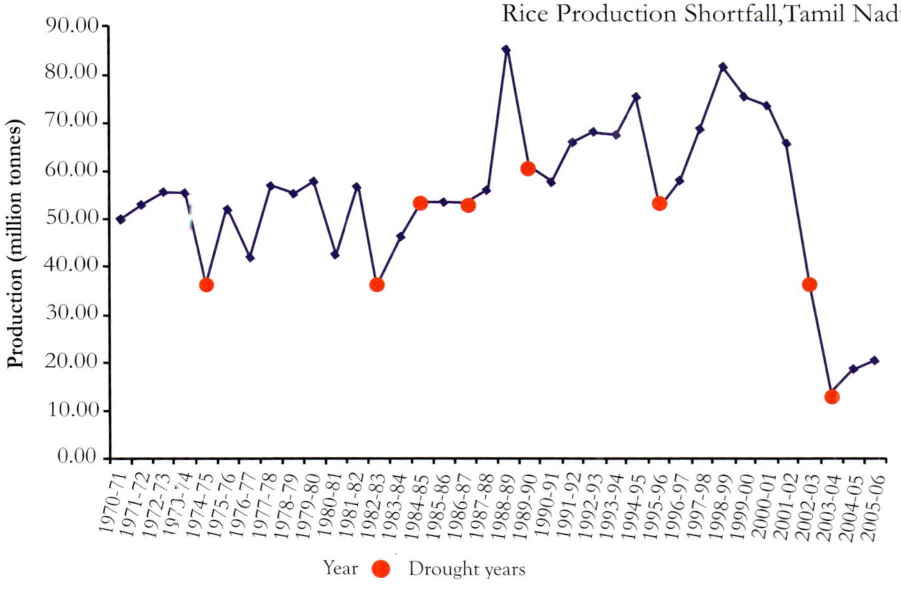

TABLE 10.
Average Rainfall during Normal and Drought Periods (mm)

Districts	Normal	Drought	Deficit	Short fall in rainfall (Percent)
RF	838.06	568.70	269.36 (32.14)	47.36
Tamil Nadu	964.98	693.75	271.23 (28.11)	39.10

(Figures in parentheses denote percentage of normal)
Source: Authors' estimates. based on Government data

Loss in production due to drought was calculated to be about 30 per cent of total rice production in Tamil Nadu and in value terms amounts to Rs. 852.11 crore, which accounts for 5.54 percent of the gross domestic product (Table 11). Loss in employment was 17 percent, which was calculated based on the employment elasticity of 0.6 (Bhalla, 1987). Average per hectare labor requirement for rice is 159 man-days, but the drought resulted in loss in employment of around 28 man-days per hectare and at the state level it works out to 60 million man-days.

TABLE 11.
Effect of Drought on Value of Production and Employment in Rice Cultivation in Tamil Nadu

Particulars	Details
Loss in quantity (lakh tonnes)	17.12
Loss in value (Rs. in crores)	852.11
Per cent of loss to GSDP by primarysector	5.54
Per cent of loss in production	29.03
Per cent of loss in employment	17.42
Loss in employment (mandays per hectare)	27.76
Loss in employment in the State (million mandays)	60.08
Cost of additional employment (Rs.in crores)	300.40

Source: Authors' Estimates based on a survey

Figure 5.

Drought Affected Rice Land in Tamil Nadu

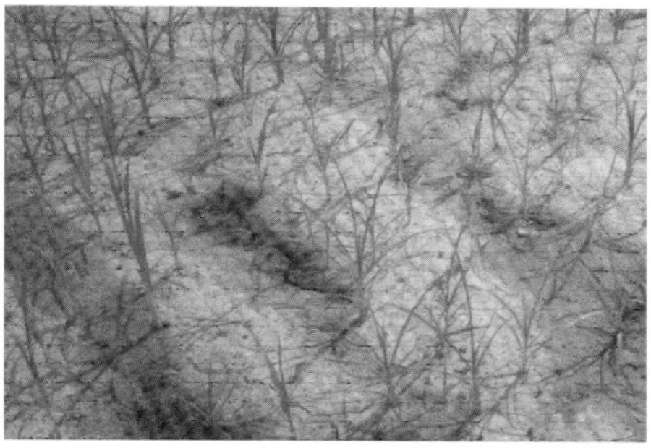

Around 50 per cent of the total irrigated area in the country is now dependent on groundwater and 60 per cent of irrigated foodp roduction depends on irrigation from groundwater wells. Declining water levels may lead to a drop in harvests in the near future. Currently, over 10 per cent of blocks, classified by the Central Ground Water Board, have been identified as 'overexploited' blocks, and the exploitation beyond the critical level has been growing at a rate of 5.5 percent per year. Further, about 36 percent of blocks in the country will be on the critical list by the year 2017. Over exploitation of groundwater resources causes either a fall in the water table, or a rise in the water table and salinity problems. Over mining of groundwater induces saline water intrusion into aquifers in coastal areas resulting in deterioration of water quality. The problem of salinity is increasing at an alarming rate due to water scarcity and over exploitation of groundwater. In India, about 8.5 million hectares are affected by salinity (Table 12), and it accounts for 4.43 per cent of the net cropped area.

TABLE 12.
Extent and Distribution of Salt Affected Soils in India (000 ha)

State	Waterlogged area			Salt-affected area			
	Canal command	Unclassified	Total	Canal command	Outside canals	Coastal	Total
Andhra Pradesh	266.4	72.6	339.2	139.4	390.6	283.3	813.3
Bihar	362.6	NA	362.6	224.0	176.0	Nil	400.0
Gujarat	172.6	311.4	484.0	540.0	327.1	302.3	1214.4
Haryana	229.8	45.4	275.2	455.0	NA	Nil	455.0
Karnataka	36.0	NA	36.0	51.4	266.6	86.0	404.0
Kerala	11.6	NA	11.6	NA	NA	26.0	26.0
Madhya Pradesh	57.0	NA	57.0	220.0	22.0	Nil	242.0
Maharashtra & Goa	6.0	105.0	111.0	446.0	NA	88.0	534.0
Orissa	196.3	NA	196.3	NA	NA	400.0	400.0
Punjab	198.6	NA	198.6	392.6	126.9	Nil	519.5
Rajasthan	179.5	168.8	348.3	138.2	983.8	Nil	1122.0
Tamil Nadu	**18.0**	**109.9**	**127.9**	**256.5**	**NA**	**83.5**	**340.0**
Uttar Pradesh	455.0	1525.6	1980.0	606.0	689.0	Nil	1295.0
West Bengal	NA	NA	NA	Nil	NA	800.0	800.0
Total	**2189**	**2338**	**4528**	**3469**	**3027**	**2069**	**8565**

Source: www.agristat.com

Rice Biotechnology Research in India

Recent advances in plant molecular biology have opened up new avenues to apply biotechnology tools for improving the conventional approaches. Increasing investment in agricultural biotechnology by both the public and private sectors is being made. India's has emphasized agricultural biotechnology because of the need to ensure food and nutrition security for current and future populations. India has one of the world's largest public sector plant breeding enterprises functioning under the overall umbrella of the Indian Council of Agricultural Research. The Indian Council of Agricultural Research (ICAR) has established a National Research Centre on Plant Biotechnology at the Indian Agricultural Research Institute (IARI) in New Delhi. Various Agricultural Universities have initiated their own biotechnology programs with funding from national and international agencies and the state governments. Three lines of molecular research are emerging from the established programs namely gene isolation, marker assisted breeding, and the production of transgenic plants for resistance to biotic and abiotic stresses.

In the Department of Science and Technology (DST), a Biotechnology Board was created in 1982. In 1986, the board was upgraded to an independent Department of Biotechnology (DBT) to give a major thrust in molecular biology and biotechnology research. Currently, much of the plant molecular biology and crop biotechnology work in India has been supported by DBT, including rice biotechnology research. The Department supports seven Centres of Plant Molecular Biology (CPMBs) in different parts of the country with the goal to support research and human resource development in plant molecular biology. They are functioning at Jawaharlal Nehru University (JNU), New Delhi, Tamil Nadu Agricultural University (TNAU), Coimbatore, Madurai Kamaraj University (MKU), Madurai, National Botanical Research Institute (NBRI), Lucknow, Osmania University (OU), Hyderabad, Bose Institute (BI), Calcutta and University of Delhi South Campus (UDSC), New Delhi. Under DST, the institutes like Indian Institute of Science (IIS), Bangalore, Centre for Cellular and Molecular Biology (CCMB), Hyderabad and the Centre for DNA Fingerprinting and Diagnostics (CDFD) at Hyderabad also contribute significantly in plant molecular biology.

Rice Biotechnology Research in TNAU

The Tamil Nadu Government has entrusted Tamil Nadu Agricultural University (TNAU) with the task of managing agricultural research at the state level with the objective of increasing rice production per unit of resource (land, water, fertilizer, and labor). TNAU carries out its mandate of rice research through various centers scattered throughout the state. These centers cater to the needs of seven agro-climatic zones of the state by stressing applied research in accordance with soil and ecological conditions of the zones. In addition, the Centre for Plant Molecular Biology at Coimbatore undertakes research work on frontier areas such as genetic engineering, marker technology, tissue and anther culture to improve rice yields. About one third of the agricultural scientists of TNAU address issues relating to rice in one way or another. Most rice research centers engage in varietal breeding, crop selection, adaptive research, resource management and plant protection.

The state agricultural research system in Tamil Nadu has a long history of research accomplishment with its formal rice research dating back to 1908. The wonder rice, ADT27, was released in 1965 marking the beginning of the green revolution in Tami Nadu. Generally, the research focus has been largely directed at irrigated rice in Tamil Nadu and very little attention has been paid to research on rain fed rice. However, in recent years, more resources are being channeled to develop rice varieties meant for rain fed systems. Rice research is funded mostly by Tamil Nadu State Government. Indian Council of Agricultural Research (ICAR) and is the second major source of funding for rice research in Tamil Nadu under the All India Coordinated Rice Improvement Programme (AICRIP). The budget from ICAR amounts to about 15 per cent of the total budget meant for rice.

Useful research outcomes and subsequent gains are not the product of a single institution. Realizing that fact, TNAU has established close linkages with IRRI and the Rockefeller Foundation for several decades. A considerable amount of germplasm and knowledge have been exchanged between scientists working in different institutes through networking. Rice research in Tamil Nadu preceded the green revolution and this has been instrumental in extending research beyond the initial success. Since 1980, scientists have given attention to developing varieties with characters that were missing earlier and were also suitable for different agro-ecosystems, incorporating suitable traits for enhancing productivity under stress conditions. (Figure 6)

New genomics and molecular breeding open up opportunities for fostering sustainable advances in crop productivity. Biotechnology holds promise for augmenting agricultural production. Research and development in biotechnology, including development, field testing and commercialization of crops, is now recognized to be an essential, and increasingly important element of a strategy for integrating both conventional and biotechnology applications to achieve future food security

Figure 6.
Rice Bio-technology Research in Tamil Nadu Agriculture University

Drought and Salinity Tolerant (DST) Rice

Breeding varieties for sustainable production under moisture stress conditions has always been difficult because selection becomes either impossible or inefficient in the absence of a representative drought period during the testing season (Blum, 1998). Drought tolerance is a complex character resulting from the interaction of many quantitative component traits. Across the globe several scientific teams have been working on genetically engineered rice and other crops to make them more tolerant to drought, salinity and temperature stresses, while improving yields. Biologists working at Cornell University, led by Prof. Ray Wu, have reported success in introducing the genes for trehalose sugar synthesis into indica rice varieties. Trehalose is a simple sugar that is produced naturally in a wide variety of organisms – from bacteria and yeasts to fungi, including mushrooms, and in many invertebrates, particularly insects. But, there is normally not much trehalose accumulated in plants, with the exception of the so-called resurrection plants that can survive prolonged droughts in deserts. Drought stressed resurrection plants look as if they are dead and gone forever, but then they spring back to life when moisture is available. Transgenic rice plants with the trehalose-enhancement gene sequences have been tested through five generations in the greenhouse – from seed producing plants to seedlings and more seed-producing plants, again and again - and the desirable, stress tolerance characteristics have been observed (Segelken, 2002). Compared with non-engineered rice plants that lack the trehalose-enhancement gene sequences, the transgenic rice plants are much more robust under a variety and combination of environmental stresses. Professor Wu's lab has also been working with several other genes that promise to improve the tolerance of rice to drought/salinity. However, the varieties that have been transformed can not be grown in other countries, so the technology must be transferred to suitable varieties before transferring it, even for field testing (Segelken, 2002). Considerable efforts have also been made over the last three decades to develop drought and saline tolerant genotypes at IRRI in collaboration with other partner institutes in India. The bottom line is that transgenic DST rice is likely to be at least 10 years away from commercialization.

Sources of Data

This study includes two sets of sample respondents: farm households and scientists. The number of blocks/taluks of the respective districts was selected for sampling based on discussions with Agricultural Department officials. Villages and farmers were also selected based on the discussions with Agricultural Department officials. Scientists were selected based on their specialization.

For collection of farm level data, 150 rice farmers were selected from Tamil Nadu and Chattisgarh. These areas are representative of drought and salt tolerant areas in other states in India as well. At present, two public sector institutions - Tamil Nadu Agricultural University and Madurai Kamaraj University and a private sector group, SPIC Science Foundation, are undertaking research on transgenic rice in the Tamil Nadu State. Indira Gandhi Agricultural University at Raipur, Chattisgarh State is presently undertaking research on transgenic rice. Further, scientists working on transgenic rice at the Directorate of Rice Research, Hyderabad were personally contacted. From these institutions, 11 scientists were interviewed to gather information pertaining to research and regulatory cost, likely adoption, and probability of research success. State-wise distribution of sample farmers and scientists is presented in Table 13. Secondary data were also collected from various published sources. These data included items such as rice production, prices, trade, price elasticities, and discount rate.

TABLE 13.
Respondents Distribution across the States

State	Targeted Problem	Selected Districts	Selected Institutes	Respondents	
				Farmers	Scientists
Chattisgarh	Drought	Raipur	IGAU-Raipur	30	2
Tamil Nadu	Drought and Salinity	Nagapattinam, Pudukkottai, Ramnad and Coimbatore	SPIC Science Foundation-Tuticorin TNAU,CBE	120	7
Andhra Pradesh	Drought and Salinity	–	DRR-Hyderabad		2
Total	–	–	–	150	11

Technology Adoption and Yield Gaps

In the rain-fed rice production environment most of the farmers practice only traditional methods of cultivation. The farmers rely mainly on traditional varieties, which are low yielding but tolerant to water stress. Land races are cultivated to a large extent in rain-fed ecosystems and farmers prefer to grow land races due to their drought tolerance. Land races generate moderately higher revenue during the drought period compared to HYVs, despite higher yield of modern varieties, due to low cost of production. If farmers perceive an improved variety to be inferior to traditional varieties in terms of one or more attributes, they are unlikely to adopt the new variety. Therefore, it is imperative that a variety meant for a water limiting environment attain at least a minimum level of yield during the stress period if farmers are going to be induced to adopt it. Yield gaps of major varieties are listed in Table 14.

TABLE 14.
Yield Gap of the Major Varieties in the Drought Prone Area (kg/ha)

Varieties	Potential Yield	YG I	YG II
J-13	4500	2300	100
ADT-43	5900	2200	600
CO-43	5200	2020	800
IR-20	5000	2940	400
ADT-39	5000	2600	500
Culture Ponni	4750	1800	200
ADT-36	4000	1300	200
Bapatla	4200	1200	800
Deluxe ponni	4900	1750	300

Yield gap of preferred varieties in drought area YG I=Yield Gap I : Potential Yield –actual Yield
YG II = Yield Gap II: Maximum Farmer's Yield-Average of Farmers' Yield
Source : Farm Survey

Any crop grown in fragile environments must be adapted to the climatic conditions, and often climate alone is responsible for large gaps between potential and actual yield (Thompson et al., 1997). Salinity effects influenced by climate, and salinity-induced yield gaps, differ across the growing seasons. Rice is a crop moderately sensitive to salinity. Varieties Co-43, ADT-37, 38, 43, ASD-16, Super Ponni and Culture Ponni were widely cultivated by the farmers. Out of the 19 varieties cultivated in the study area, saline and alkaline tolerant varieties TRY-1, ASD-16 and Co-43 were released by Tamil Nadu Agricultural University (TNAU).

About 50 per cent of the rice cultivators adopted these partially saline tolerant rice varieties and in terms of area coverage, these varieties accounted for 25 per cent of the total area under rice production in the sample farms. The majority of the farmers (38.40 per cent) in saline affected areas adopted these varieties and ADT 37, 38 and 43 and Ponni. These latter varieties were popular among the farmers due to grain quality and consumer preference despite the fact that the yield gap was higher for them (Table 15). Yield gap I in Table 15 implies that a major part of the potential yield was left untapped even in the case of most progressive farmers. To a large extent this yield gap is attributable to environmental disparities such as soil salinity. As a result, the technology developed at the research station could not be fully replicated in the farmers' fields. Variety-wise yield gap analysis showed that yield gap I was higher (2633 kg per ha) in the case of Co47, while it was less (1250 kg per ha) in the case of Ponni. Among the cultivars, yield gap II was found higher (611 kg per ha) in the case of CO 43

TABLE 15.
Yield Gap of the Varieties in the Salinity Prone Area (Kg/Ha)

Varieties	Potential Yield	YG I	YG II
ADT 36	5000	1500	428
ADT 37	5300	1831	371
ADT 38	5800	2613	120
ADT 39	5000	1750	159
ADT 43	5900	2150	610
Ponni	5000	1250	594
CO 43	5200	1450	611
CO 47	5800	2633	76
Deluxe ponni	4900	1750	300

Yield gap of preferred varieties in salinity area YG I=Yield Gap I: Potential Yield –actual Yield
YG II = Yield Gap II: Maximum Farmer's Yield- Average of Farmers' Yield
Source : Farm Survey

Data Analysis

As discussed in chapter 1, budgets were constructed and economic surplus changes due to adoption of DST rice were calculated. These benefits were compared to research and regulatory costs and net present values were calculated. In the economic surplus analysis, the elasticity of supply for rice was assumed to be .40 and elasticity of demand, -.35. The probability of research success was assumed to be .68 based on interviews with scientists. A real discount rate of 5 percent was assumed. Costs and benefits were estimated over a 17 year period. Internal rates of return and net present values were then calculated.

Results of comparing (DST) Transgenic and Non-Transgenic Rice

Adoption of transgenic drought and salinity tolerant (DST) rice is projected to bring additional income to farmers, despite an increase in seed cost. The cost of seed for DST rice is projected to be about 15.5 percent higher than for the existing high yielding varieties, and the level of use of other inputs should remain about the same. Yield of DST rice would be 25 percent higher as compared to existing varieties under stress conditions. Labour is the major input accounting for 28 percent of the total cost of production followed by fertilizer (35 percent) and seed (15 percent). Farmers would incur Rs. 7998 per hectare in cost of cultivation, while the total return would be Rs. 26114 per hectare. Partial budgeting was used to estimate additional costs and returns from growing one hectare of DST rice in place of existing high yielding varieties. Adoption of transgenic drought and saline tolerant rice would bring an additional 25 percent in income per hectare despite the increased cost of seeds (Table 16 and Figure 7).

Return on Investment

The estimated internal rate of return (IRR) for transgenic rice was calculated to range from 161 to 206 percent in several scenarios, whether trade was assumed or not (Table 17). The Net Present Value (NPV) over 17 years was estimated in dollar terms to range from 1028 to 3343 million US$. These results indicate that investment in DST rice brings high economic and social benefits, suggesting a continued need for research investment in transgenic rice. As rice production forms the major source of income in these water limiting and fragile environments, considerable potential exists for reducing poverty due to realization of higher levels of income by adopting transgenic technology. Since genetic engineering could improve the crops resistance to water stress and soil adversity like salinity, there would likely be a very high rate of adoption for this new technology among the farmers in the years to come. In India a large-scale application of this technology is expected.

TABLE 16.
Costs and Returns of Transgenic (DST) and Non-Transgenic Rice (Rs/ha)

Cost	Non-Transgenic	Transgenic	Difference (%)
Seed	1046.25	1208.42	(+)15.5
	(13.34)	(15.11)	
Labour	2240.00	2240.00	–
	(28.59)	(28.00)	
Bullock Labour	641.68	641.68	–
	(8.19)	(8.01)	
Machine Power	766.65	766.65	–
	(9.77)	(9.59)	
Fertilizer			
FYM	1202.5	1202.5	–
	(15.35)	(15.02)	
Urea	585.43	585.43	–
	(7.46)	(7.32)	
Single super phosphate	224	224	–
	(2.86)	(2.80)	
Diammonium phosphate	485.68	485.68	–
	(6.20)	(6.06)	
Potash	277.93	277.93	–
	(3.55)	(3.46)	
Pesticides			
Insecticides	182.65	182.65	–
	(2.32)	(2.27)	
Fungicides	183.33	183.33	–
	(2.34)	(2.28)	
	365.98	365.98	–
	(4.66)	(4.55)	
Total Cost	7836.10	7998.27	–
	(100.00)	(100.00)	
Return			
Yield (tonnes)*	3.73	4.68	(+)25.00
Price (Rs/kg)	5.58	6.70	(+)20.00
Gross Income (Rs/ha)	20813.4	26114.4	(+)25.46
Net Income (Rs/ha)	12977.3	18116.13	(+)39.60
Cost Benefit ratio	2.66	3.27	–

5 (Maximum realizable)
Predominant varieties are – Swarna, Mahamaya, Poornima, IR 64, Deluxe Ponni, Culture Ponni and Co 43
Source: Farm Survey

Conclusion

Drought is a common recurring event in Indian agriculture with a large effect on rain fed rice production and the livelihoods of farmers. Intensive irrigated agriculture as well as coastal water incursions has led a serious salinity problem that reduces rice production and farm income. Therefore, development and commercialization of DST rice is potentially beneficial and our results show that adoption of transgenic drought and salinity tolerant (DST) rice is projected to bring increased net revenue to farmers, despite an increase in seed cost. Total economic benefits for India from DST rice are projected to range from 1028 to 3343 million US $ over 17 years.

Reference

Batia, M.S. (1999). 'Rural Infrastructure and Growth in Agriculture', *Economic and Political Weekly*, Vol 34: pp A 43-A 68.

Bhalla S. (1987). Trends in Employment in Indian Agriculture, Land and Asset Distribution. *Indian Journal of Agricultural Economics*, 42(4):537-560.

Blum A,(1998)Drought Resistance. in Plant Breeding for Stress Environments. CRC Press, Boca Raton, Florida, pp 43-76,

Hazell, P.B. and C.Ramasamy (1991): The Green Revolution Reconsidered: The Impact of High Yielding Rice Varieties in South India, Johns Hopkins University Press, Baltimore, MD.

Mellor JW (1978): Food Price Policy and Income Distribution in Low Income Countries: *Economic Development and Cultural Change*.27:1-26

Mishra, B, (2005): "More Crop Per Drop", The Hindu Survey of Indian Agriculture. p. 41-46.

National Commission of Agriculture (1976), Government of India, New Delhi.

Pingali, P.L, M.Hossain and R.Garpacio (1997): Asian Rice Bowls: The Returning Crisis? CAB International Rice Research Institute, Philippines.

Pingali, P.L. and M.Hossain (eds) (1999): Impact of Rice Research in Asia, International Rice Research Institute, Philippines.

Segelken. R, (2002): Stress Relief: Engineering Rice Plants with Sugar-producing Gene Helps Them Tolerate Drought, Salt and Low Temperatures, Cornell Biologists Report, Cornell News. p. 30.

Siddiq, E.A., (2000): Rice: Yawning Productivity Gaps. Survey of Indian Agriculture, The Hindu, P.39

Thompson J, I. Hume, D. Griffin, S. North and D. Mitchell. (1997): Use of Saline Water in Rice Based Farming Systems, National Program for Irrigation Research and Development, NSW Agriculture, Deniliquin, NSW 2710.

www. Agristat.com
www.fao.org

Chapter **3**

Tobacco Streak Virus Resistant in Groundnut and Sunflower

K.N. Selvaraj, C. Ramasamy and G.W. Norton

Introduction

Yields are low in developing countries for many reasons, but one of the principal causes is that crops suffer more from biotic stresses due to insects, weeds, and diseases than they do in temperate climates. In India, research on transgenic plants has been in progress in the public and private sectors to develop solutions to some of these pest problems, including a virus problem on groundnuts and sunflower. Research and development in biotechnology, including development, field testing and commercialization of crops, is recognized as an increasingly important component of a strategy to achieve food security. Breeding varieties for resistance/tolerance to biotic and abiotic stresses as well as for improved nutritional qualities are necessary to meet challenges rising from population growth, environment degradation and nutritional insecurity. During past three decades, improvement of major crops has been achieved mainly through classical genetics and plant breeding methods. Public and private organizations are now funding agricultural biotechnology, including development of transgenic plants, to add tools to the arsenal. This chapter reports on analyses to evaluate potential impacts of transgenic research to develop groundnut and sunflower varieties with resistance to tobacco streak virus, a major yield-reducing virus.

The Oilseed Sector in India

India is one of the largest producers of oilseeds in the world. There has been a significant increase in oilseed production since the mid 1980s and the increase can be attributed to both area and productivity growth, the latter resulting from improved technologies. Production increased from 5.90 million tonnes in the 1950s to 21.35 million tonnes in 2004-05. Groundnut is the major oilseed accounting for more than 40 per cent of the area under oilseeds, but its share has declined over the years due to increases in soybeans and sunflower (Table 1). Rapeseed and mustard are the next most important oilseeds in terms of acreage, accounting for 30 per cent of the oilseeds produced in India. Oilseed yield almost doubled over the last four decades.

TABLE 1.
Estimated Internal Rates of Return and Net Present Values (NPV) of DST Rice

Commodities	Area		Production	
	970-71 to 1990-91	1990-91 to 2003-04	1970-71 to 1990-91	1990-91 to 2003-04
Groundnut	1.65	-1.88	2.97	3.92
Sunflower	4.07*	-10.60***	6.00**	-12.90***
Oil seeds	2.33***	-3.45***	4.05***	-5.12

* Significant at 10 per cent level
** Significant at 5 per cent level
*** Significant at 1 per cent level

The share of oilseeds in total cropped area increased only marginally over the last three decades. Only 10 per cent of the area under oilseeds is irrigated and as a result yields tend to be low. There is a diversion of major oilseeds such as groundnut, rapeseed, and mustard from irrigated to non-irrigated conditions. Oilseeds are grown on marginal land holdings under natural climate and low fertility conditions (Ramasamy and Selvaraj, 2002). Evidence indicates that area variability in oilseeds is high (Nagaraj and Gowda, 1997; Jhala, 1997) which leads to production variability, resulting in a surplus in one year and a deficit the next. Groundnut yield was 2168 kg per hectare in irrigated areas as compared to 1414 kg per hectare in rain-fed areas, a gap of 754 kg, or almost 50 per cent of the irrigated yield (Ramasamy and Selvaraj, 2002). The average oilseed yield in the country is around 0.68 tonnes per ha, far below the world average of 1.43 tonnes per ha and the average of 1.63 tonnes (Table 2) per ha achieved in China during the same period (Pandey and Kandulna, 2000).

TABLE 2.
Productivity of Oilseeds in India and the World

Crops	Average yield	Highest yield	World average yield (kg/ha)	India' s average yield to world's average yield (percent)	India's highest yield to world's average yield (percent)
Sun flower	549	1429	1225	44.82	116.65
Groundnut	977	1471	1367	71.47	107.61

Trade in Oils

Recently, reduction and rationalization of tariffs and removal of non-tariff barriers have played a crucial role in increasing imports in India. A major import component is food and related items, of which pulses and edible oils are major share. The country imports about 15 to 18 lakh tones of edible oil each year (Business Line, 2000). During 2003-04 edible oil imports rose to more than $2.54 billion. This raised Indian's dependence on imports for edible oil to almost 40 percent (Chand and Pal, 2003), causing an adverse impact on domestic oilseed growers (Chand, et al., 2003).

In the post reform period, the growth in area and production was negative in most of the oilseeds. Imports of edible oil have increased 933 percent as compared to pre reform period. Edible oil exporting countries like Malaysia, Indonesia, and Brazil are beginning to flood the Indian market with palm and soybean oil. The edible oil industry feels that the dumping of palm oil and other oils into the country has hampered the development of oilseed cultivation. Any increase in production of oilseeds will have to be contributed mainly by an increase in average productivity of oilseeds.

Technological Constraints

Failure to achieve higher yields in oilseeds is one of the reasons for the widening gap between domestic production of vegetable oils and demand. This is due to insufficient technological improvements, especially high yielding seed material and plant protection technologies. In the case of oilseeds, more than 300 modern varieties and hybrids crops have been released to the farmers (www.icar.org) but their adoption rates are poor due to their cultivation in rain-fed areas, which are endowed with poor resources. In rain-fed area, farmers are reluctant to adopt improved technology due to low returns. A study on adoption of various technology components for select oilseed crops reported that 56 percent of farmers adopted a full complement of technologies; 37 percent adopted partially, and 8 percent adopted none. There exists a realizable yield gap of about 7.5 quintal per hectare (Kiresur et al., 2001) due to poor adoption of modern technologies. Shortage of improved quality seeds, high cost of oilseeds, and the low priority accorded to oilseeds research and development resulted in a lack of response by farmers. If attempts were made to increase the yield of 10 traditional oilseeds, and those with high oil-recovery, by 20 percent, the resultant increase in edible oil supplies would be at least 15-lakh tonnes. This increase would be sufficient to meet the growing demand.

As oilseeds are mostly grown in rain-fed areas, farmers are reluctant to use quality inputs, such as chemical fertilizers and pesticides. Farmers mainly use their own seeds, untreated and worn-out varieties. Adoption of improved technologies is not taken seriously under rain-fed conditions. Therefore, increasing oilseed production will be tough unless one resorts to improved technology particularly varieties suitable to various production environments with resistance to biotic and abiotic stresses. This technology problem can be looked at from two angles: one is whether desired technologies are available and the other is whether improved technologies for oilseeds crops are superior in terms of yield and profit. Many studies arrive at the conclusion that there is a substantial gap between potential and actual yields, and if a recommended package of technology and cultural practices is followed by farmers, it may not too difficult to bridge this gap. The economic viability of technologies is not the constraint, but poor adoption of improved technologies.

Tobacco Streak Virus

Groundnut and sunflower are prone to attack by various diseases and insects. Among the foliar fungal diseases, leaf spot (early and late) is economically important. Among the viral diseases, Bud Necrosis (BND), Peanut Mottle (PMV) and Peanut Clumb (PCV) are economically important. An attempt was made to understand the likely adoption of Tobacco Streak Virus resistant transgenic varieties of sunflower among the growers in India.

Recently the tobacco streak virus caused significant damage to oilseed crops, particularly groundnut and sunflower (Figures 1 and 2).

Figure 1.
TSV (Tobacco Streak Virus) Affected Groundnut Field

Figure 2.
Tobacco Streak Virus Affected Sunflower Plant

A disease epidemic resulted in the death of young groundnut plants. The disease occurred in the rainy season in 2000 in the Ananthapur district, Andhra Pradesh, where the crop is grown usually on 0.7 million hectares. The disease affected nearly 255,000 ha and crop losses were estimated to exceed Rs 3 billion (US$65 million). Another epidemic occurred in the rainy season of the year 2003. Characteristic symptoms of the disease are necrosis of the stem and terminal leaflets followed by death of the plant. Severe yield reduction is noticed from all infected plants and the majority of pods show necrotic spots. Scientists were of the opinion that initially, the disease was caused by the tospovirus Peanut bud necrosis virus (PBNV) because of the characteristic necrosis of terminal leaflets. Axillary shoot proliferation and severe leaf deformity are characteristic symptoms of peanut bud necrosis disease (PBND). During the field survey it was observed that many of the affected peanut fields were adjacent to sunflower fields, which showed the symptoms of sunflower necrosis disease. The most economical way to manage PSND, as for most other plant virus diseases, is to grow resistant varieties.

The appearance of Sunflower Necrosis Disease (SND) was observed for the first time during 1997 at Bagepalley near Bangalore (Karnataka) which later spread to the neighboring states (Andhra Pradesh, Karnataka, Tamil Nadu and Maharashtra) posing a serious threat to sunflower cultivation. Surveys conducted from 1998 through 2001 revealed that the disease was widespread and its intensity levels ranged from 2-100 percent. The disease caused severe epidemics in some places resulting into total yield loss, which made the farmers abandon sunflowers and switch over to other crops. Scientists believed that the most economical way to manage PSND, was to grow resistant varieties. But resistance to TSV has not yet been found in cultivated groundnut or sunflowers. All the released varieties of groundnut in India were susceptible. Removal of weeds helps in reducing disease incidence, but removal of infected groundnut plants from the field has no effect. During field surveys, soon after the TSV outbreak in Ananthapur, natural barriers such as tall grass in the fields were found to protect the adjacent crops from the disease. In one case with a tall grass barrier between a groundnut field and parthenium in the adjacent field, the crop was free from the disease, whereas a nearby groundnut crop without a barrier showed high incidence of streak virus. The scientists believe that naturally growing grass might have obstructed the wind borne thrips and the inoculum carrying pollen of parthenium from landing on groundnut plants (Figure 3).

Other suggested practices are growing 7-10 rows of fast growing (tall), pearl millet, sorghum or maize as border crops around groundnut fields. Farmers are advised to maintain normal plant population because sub-optimum plant population leaves bare patches in the field, which attract thrips. It is advisable not to grow sunflower, marigold and other TSV susceptible crops adjacent to groundnut fields. As is the case with most of the thrips-transmitted viruses, use of insecticides after the appearance of disease has no effect on disease control. Seed treatment is also recommended.

Figure 3.
Parthenium Sps. - Host for vector of Tobacco Streak Virus

In the case of sunflower, as the disease is of a viral nature, emphasis has been on controlling the insect vector through chemicals. However, under high disease severity, chemical protection has limited effect. Other practices suggested are removing weeds particularly parthenium from the field and adjoining areas of crop, avoiding chrysanthemum and marigold close to sunflower, crop rotation, avoiding highly susceptible cultivars, growing 5-7 rows of a border crop such as sorghum or pearl millet, seed treatment, and uprooting and destroying disease-affected plants when observed.

Data Collection

Groundnut is mainly produced in the States of Andhra Pradesh, Tamil Nadu, Gujarat, Karnataka and Maharashtra. These four states account for nearly 90 per cent of groundnut production in the country. Therefore, 80 groundnut and 30 sunflower growers in Ananthapur district of Andhra Pradesh, Bangalore Rural and Kolar districts of Karnataka, and Junagadh district of Gujarat were chosen to be personally interviewed with pre-tested questionnaire (Tables 3 and 4). Scientists were also selected based on their transgenic specialization in the target crop. Sixteen scientists (13 for groundnut and 3 for sunflower) working in State Agricultural Universities, ICAR institutes, and private sectors across the state were personally interviewed (Table 5). The scientists helped with the farm level surveys conducted in each state. Villages and farmers were selected based on the discussions with the Agricultural Department Officials and data were collected using random sampling (Figure 4).

Figure 4.
Field Survey Research Investigators in the TSV Affected Farmer's Field in Andhra Pradesh, India

TABLE 3.
Respondent Distribution across the States in India

State	Respondents	
	Farmers	Scientists
Andhra Pradesh	30	9
Karnataka	50	5
Gujarat	30	2
Total	**110**	**16**

TABLE 4.
Respondent Distribution across Scientists and Crops

Crops	Respondents	
	Farmers	Scientists
Groundnut	80	13
Sunflower	30	3
Total	**110**	**16**

TABLE 5.
Scientist Distribution across Institutes

Crop	Number of Scientists	Place of working
Groundnut	5	UAS – Bangalore
	4	ICRISAT & NBPGR – Hyderabad
	2	IARI – New Delhi
	2	NRCG – Junagadh
Sunflower	3	DOR, Hyderabad
Total	**16**	

Data were gathered on yield losses, practices for managing the target pests, and crop yields and costs. Experimental data and opinions of scientists were used in addition to input from farmers and secondary data. Data were gathered on time and cost of research, price elasticities of demand and supply, output and output prices, trade, and pesticide consumption. Two sets of interview schedules were prepared for scientists and households and pre-tested. Respondents were contacted individually and the objectives of the study were clearly explained to ensure co-operation, sincerity and accuracy in their responses and information. The primary information pertaining to area of specialization, scientists' perceptions on preferred varieties by the farmers and probable solutions for their problems, expected change in the consumption of inputs due to adoption of DST technology, cost of technology development, environmental benefits, limitations and remedies for transgenic crop research were also collected. Data collected were coded, processed and classified into tables in order to bring out results that could be generalized.

Data Analysis

As discussed in chapter 1, budgets were constructed and economic surplus changes due to adoption of TSVR groundnut and sunflowers were calculated. These benefits were compared to research and regulatory costs and net present values calculated for both groundnut and sunflower. A Technology Adoption Index was also estimated.

Results

Adoption of Management Practices

A gap exists between actual yield of the crop realized by the farmers in their field and the yield that could be obtained. Therefore a Technology Adoption Index (TAI) was constructed to categorize the farmers as low, medium and high adopters of various management practices. In constructing the adoption index, various components were included. The adoption index ranged between 0 to 100 percent. Farmers having less than 40 percent of the TAI were grouped as low adopters, farmers with a TAI between 40 to 70 percent were classified as medium adopters, and farmers with more than 70 per cent of the TAI were grouped as high adopters. Results indicate that the most of the farmers are not adopting the suggested practices such as seed treatment, and operations such as removing parthenium weeds (Table 6). Most of the farmers resorted to chemical control. Farmers believed that most of the varieties are susceptible to this disease and that sometimes even chemical control is not effective, resulting in yield loss. When they were asked about whether they would prefer to grow resistant bio-engineered varieties, most of them said yes.

TABLE 6.
Adoption of Various Technological Packages (Number of Farmers)

Particulars	Andhra Pradesh	Karnataka	Gujarat	Total
Low adopters	22	40	19	81
	(73.33)	(80.00)	(63.33)	(73.64)
Medium adopters	6	4	8	18
	(20.00)	(8.00)	(26.67)	(16.36)
High adopters	2	6	3	11
	(6.67)	(12.00)	(10.00)	(10.00)
Total samples	30	50	30	110
	(100.00)	(100.00)	(100.00)	(100.00)

(Figures in parentheses are percentages of farmers)

Costs and Returns

Budgets were constructed for non-transgenic and transgenic oilseeds. Major inputs were labour, seeds, fertilizers, manures and agrochemicals. Scientists were of the opinion that seed cost would be 20 percent higher than the existing high yielding varieties and the use of other inputs would remain the same for TSVR groundnut and sunflower as it is for existing varieties.

Adoption of transgenic TSVR groundnut would bring 90 percent higher profits to the farmers despite an increase in the cost of seeds of 20 per cent compared to existing varieties. There would be an 8 percent reduction in labour use due to a reduction in application of fungicides. Farmers would incur Rs. 3573 per hectare in cost of cultivation of TSVR groundnut, compared to Rs. 3651 for existing varieties. Total returns would be Rs. 38556 per hectare in case of TSVR groundnut, compared to Rs. 26792 for existing varieties (Table 7 and Figure 5). Similarly, adoption of transgenic TSVR sunflower would bring 150 percent higher profits per hectare despite increase in seed cost of 20 percent. There would be reduction of labour use of 9 per cent due to 50 per cent reduction in application of fungicides as compared existing varieties. Yield of TSVR sunflower would be 20 percent higher than existing varieties such as Ganga Cauvery, Kargil and Suntech 120. Though there is no perceptible reduction in cost of cultivation by adopting TSVR sunflower, the benefits realized by the farmers would be higher due to the yield increase (Table 8 and Figure 6).

TABLE 7.
Costs and Returns of Transgenic (TSVR) and Non-Transgenic Groundnuts (Rs/ha)

Cost	Non-Transgenic	Transgenic	Difference (%)
Seed	2200 (16.12)	2640 (19.44)	(+) 20
Labour	4080 (29.89)	3780 (27.84)	(-7.94)
Bullock Labour	3600 (26.36)	3600 (26.53)	–
Fertilizer			
FYM	2127.5 (15.59)	2127.5 (15.68)	–
Urea	250 (1.82)	250 (1.83)	–
Single Phosphate	510 (3.74)	510 (3.76)	–
Potash	380 (2.77)	380 (2.80)	–
20:20 complex	285 (2.07)	285 (2.10)	–
Pesticides			
Insecticides (kg)	218 (1.60)	–	–
Fungicides (kg)	–	–	–
Total Cost	13650.5 (100)	13572.5 (100)	–
Return			
Yield (tonnes)	1.7	2.04	(+) 20
Price (Rs/kg)	15.76	18.90	(+)
Gross Income (Rs/ha)	26792	38556	(+) 43.90
Net Income (Rs/ha)	13141.5	4983.52	(+)
Cost Benefit ratio	1.95	2.83	–

* Predominant varieties are – TMV2, JL 24

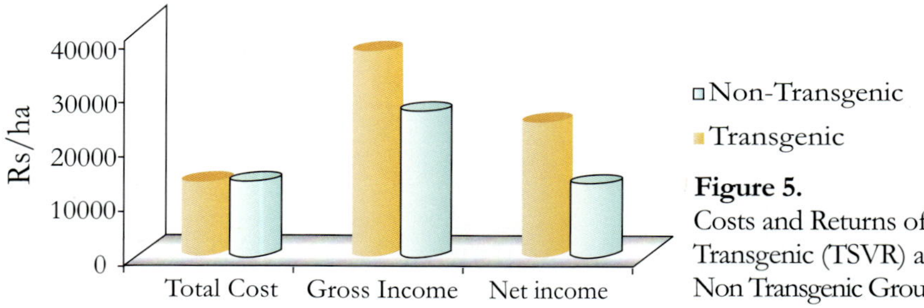

Figure 5.
Costs and Returns of Transgenic (TSVR) and Non Transgenic Groundnut

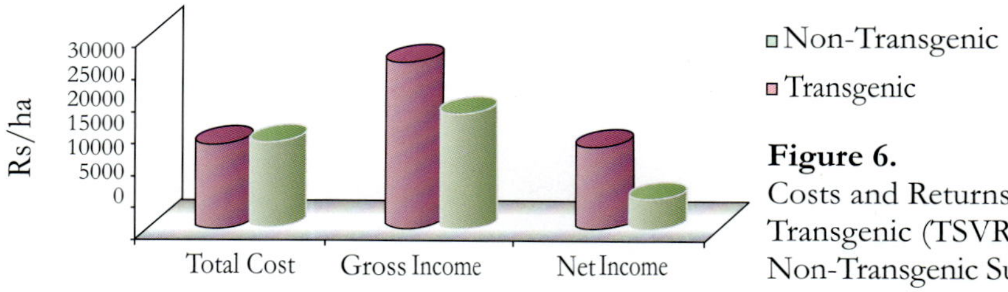

Figure 6.
Costs and Returns of Transgenic (TSVR) and Non-Transgenic Sunflower

TABLE 8.
Cost and Returns of Transgenic (TSVR) and Non-Transgenic Sunflower (Rs/ha)

Cost	Non-Transgenic	Transgenic	Difference (%)
Seed	1572.5 (11.75)	1887 (14.47)	(+)20
Labour	4500 (33.61)	4080 (31.31)	(-9.33)
Bullock Labour	375 (2.8)	375 (2.88)	
Machine power	2139.58 (15.99)	2139.58 (16.42)	
FYM	1481.25 (11.07)	1481.25 (11.37)	
Urea	412.5 (3.07)	412.5 (3.17)	
20:20 complex	44.55 (4.07)	544.55 (4.18)	
19 all complex	825 (6.15)	825 (6.32)	
17:17complex	268.75 (2.00)	268.75 (2.05)	
Diammonium Phosphate	432.5 (3.22)	432.5 (3.32)	
SSP	279.15 (2.09)	279.15 (2.13)	
Potash	58.33 (0.44)	58.33 (0.45)	
Pesticides			
Insecticides	300 (2.23)	150 (1.14)	(-50)
Fungicides	195 (1.46)	97.5 (0.75)	(-50)
Total Cost	13384.11 (100)	13031.11 (100)	(-2.64)
Return			
Yield (tonnes)	1.15	1.38	(+)20
Price (Rs/kg)	15.76	18.90	(+) 20
Gross Income (Rs)	18124.00	26082	(+) 43.90
Net Income (Rs)	4739.89	13050.89	(+) 175.33
Cost Benefit ratio	1.34	2.00	-

Predominant varieties are – Ganga Cauvery, Kargil, Suntech 120

Return on Investment

The results of the economic surplus and benefit cost analyses assuming open and closed economy models are presented in the Tables 9 and 10. Estimated IRRs for groundnut ranged from 150 to 190 percent, while it ranged from 83 to 108 percent for sunflower in the open economy model. Similarly, IRRs ranged from 136 to 174 percent for groundnut and 62 to 87 percent for sunflower in the closed economy model. Estimated NPVs over 17 years under open trade for groundnut ranged from 865 to 2688 million US$. In the case of sunflower, they ranged from 78 million US$ to 214 million US $. Under various scenarios without trade, the estimates were roughly half that. Results indicate that investing TSVR groundnut and sunflower brings high economic benefits. Such investment could also reduce India's dependence on imports of edible oils.

TABLE 9.
Estimated NPVs and IRRs for TSVR Transgenic Research-Open Economy Model

Scenarios	Groundnut NVP (Rs in million)	Groundnut IRR (Percent)	Sunflower NPV (Rs. in million)	Sunflower IRR (Percent)
Relatively Inelastic supply (**e=0.4)**	107518 (2688)	190	8544 (214)	108
Unit elastic (**e=1)**	41881 (1047)	156	3649 (91)	87
Relatively elastic **(e=1.2)**	34588 (865)	150	3105 (78)	83

(Figures in parentheses are in millions of US dollars)

TABLE 10.
Estimated NPVs and IRRs for TSVR Transgenic Research - Closed Economy

Scenarios	Groundnut NVP (Rs in million)	Groundnut IRR (Percent)	Sunflower NVP (Rs in million)	Sunflower IRR (Percent)
Relatively Inelastic supply (**e=0.4)**	69359 (1734)	174	3732 (93)	87
Unit elastic (**e=1)**	26924 (673)	142	1369 (34)	66
Relatively elastic (**e=1.2)**	17377 (434)	136	1107 (28)	62

(Figures in parentheses are in million of US dollars)

In order to assess the extent of adoption of TSVR oilseeds for the above analyses, scientists' opinions were solicited. The transgenic technology ultimately provides benefits to the farming community in terms of productivity enhancement and reduction in cost of cultivation. Since it requires a large investment; scientists favored a public private partnership in transgenic research. Further, the scientists believe that widespread application of conventional agriculture technologies such as use of herbicides, pesticides and fertilizers has resulted in severe environmental damage in many parts of India. Thus the environmental risks of new GM technologies need to be considered in the risks of continuing to use conventional technologies and other commonly used farming techniques. Most of the environmental concerns about GM technology in plants have derived from the possibility of gene flow to close relatives of transgenic plant, the possible undesirable effects of the exotic genes or traits, and the possible effect on non-target organisms. As with the development of any new technology, a careful approach is warranted before developing a commercial product. There is a need for a thorough risk assessment of likely consequences at an early stage in the development of all transgenic plant varieties, as well as for a monitoring system to evaluate these risks in subsequent field tests and releases. Scientists' believe the likely impact of cultivating transgenic oilseeds on the environment would be minimal. They were of the opinion that cultivation of TSVR groundnut and sunflower would bring benefits in terms of high yield, low production cost and high income.

Conclusion

Most of the farmers currently resort to chemical control of TSV disease. They believe that most of the varieties are susceptible to TSV disease and sometimes even chemical control is not effective in controlling yield loss. When they were asked about whether they would prefer to grow resistant bio-engineered varieties, most of them said yes. Further, results of the study reveal that adoption of transgenic TSVR groundnut would bring additional profits to the farmers in the neighborhood of 90 percent more per hectare, despite increased seed costs of 20 percent compared to existing varieties. Adoption of transgenic TSVR sunflower would bring added profits to the farmers despite an increase in seed cost. However, results of the economic surplus analysis indicate that investing in developing TSVR groundnut would bring high economic benefits to society, but returns to TSVR sunflower would be marginal.

Reference

Business Line, (2000). Daily News Paper

Chand Ramesh and Suresh Pal (2003). Technological and Policy Options to Deal with Imbalances in Indian Agriculture, *Current Science*, Special Issue, February, 2003.

Chand Ramesh, P. Kumar and Sapna Sinha (2003). Impact of Agricultural Trade and Related Reforms on Domestic Food Security in India, Report of the study done for FAO Rome, Institute of Economic Growth, Delhi, November.

Jhala, M.L. (1997). Farmers Response to Economic Incentives: An Analysis of Inter-regional Groundnut Supply Response in India. *Indian Journal of Agricultural Economics,* Vol 34 (1): pp 55-67

Kiresur, V.R; S.V.Ramana Rao and D.M.Hegde.(2001). Improved Technologies in Oilseeds Production – An Assessment of their Economic Potential in India. *Agricultural Economics Research Review.* Vol 14 (2): pp 95 – 108

Nagaraj, C.V. and M.V. Srinivasa Gowda. (1997). Growth and Instability of Safflower Production in Karnataka. *Agricultural Situation in India,* Vol. 54 (5). Pp 257-260

Pandey, R.K. and Rashmi Kandulna. (2000). Production, Consumption and Marketing of Oilseeds in India. The Bihar Journal of Agricultural Marketing. Vol 8 (4): pp 389-399

Ramasamy, C. and K.N.Selvaraj. (2002). Pulses, Oilseeds and Coarse Cereals: Why They Are Slow Growth Crops. *Indian Journal of Agricultural Economics.* Vol 57 (3): pp 289-315

www.ICAR.org

Chapter **4**

Late Blight Resistant Potato in India

K.N. Selvaraj, C. Ramasamy and G.W. Norton

Introduction

Biotechnology promises potential economic benefits to farmers in the form of lower costs for pest management, higher yield, reduced pest risk, and management time savings. Studies have shown that when disease pressure and /or weed concentration is high, specific GM crops outperform conventional crops in terms of reduced cost and additional economic benefits (Flannery, et al., 2004). If the new technology provides significant economic benefits to the producers, one might expect a high rate of adoption. Therefore investment in agricultural biotechnology in India is increasing, both by the public and the private sectors. India's emphasis on agricultural biotechnology is designed to ensure food and nutrition security for current and future populations.

Potato is an important crop in India, and late blight is a recurring phenomenon that causes significant losses, especially in the hilly regions. The disease is common in the northeastern hills where weather conditions are conducive for severe outbreaks early in the season and it remains a problem over much of the season. It can affect the entire crop before it completes its full growth. Therefore, resistant varieties are needed to combat the disease. This study reports on an assessment of the economic and environmental benefits of developing and commercializing late blight resistant (LBR) potato in India.

Potato Production in India

Potato in India is cultivated on 1.28 million hectares and accounts for about 0.65 percent of total cropped area of the country. It is grown in India in almost every state and under very diverse conditions. Nearly 80 percent of potatoes are grown in the vast Indo-Gangetic plains of North India during the short winter days from October to March. About 8 percent of the area under potato cultivation lies in the hills where the crop is grown during long summer days of April to October. Plateau regions of south-eastern, central, and peninsular India constitute about 6 percent of the area where potatoes are grown as a rain fed kharif crop during the rainy season (July to October) or as irrigated rabi crop during winter (November to March). In a small area of about 4,000ha covering mainly the Nilgiri and Palani hills of Tamil Nadu, the crop is grown round the year both as an irrigated and a rain fed crop. The states of Uttar Pradesh, West Bengal and Bihar account for nearly 73 percent of the area under potato and 80 percent of total production. There is potential for exports of processed potato and India has the potential of becoming major exporter of both ware and seed potatoes. Currently only 0.5 percent of the potato production of the country is processed. More than 80 percent of Indian potato is grown in the winter months when there is no potato crop in the Western temperate countries. India, therefore, is favourably placed as far as potato production and marketing is concerned.

Production Losses Due to Late Blight

In part due to research and development efforts, potato production in India has increased significantly and India is now the world's fourth largest producer in terms of both production and area. Productivity has increased over time due to introduction of high yielding and pest resistant varieties. Despite this fact, productivity per hectare is low as compared to other major potato growing countries in Europe, Oceana, and the United States. Low productivity can be attributed to poor seed quality, inappropriate varieties, and inadequate control of insects and diseases particularly late blight (caused by the fungus *Phytophthora infestans*). As all existing high yielding varieties are susceptible to this disease (Figure 1). Yield losses due to late blight vary from year to year (Figure 2), ranging from 19-65 percent in eastern hills, 11-74 percent in Northwestern hills, 10-75 percent in Eastern plains, 20-40 percent in northwestern plains, and 31-39 percent in southern hills (Tables 1 and 2). Late blight and bacterial wilt cause enormous yield losses of up to 75 to 80 percent in a single season. An improved variety developed by the scientists at the Central Potato Research Institute (CPRI), Shimla, was released for commercial cultivation with the name `*Kufri giriraj*' in 1999, and it has been well accepted by growers in the hilly regions of Tamil Nadu. High yielding potato with built-in resistance to late blight is widely grown in the hills of Tamil Nadu, replacing high yielding varieties such as *Kufri jyoti* and *Kufri swarna*. However, yield losses even for "resistant" varieties are high (Table 3 and Figure 3) with chemical control. In some of the States the yield loss due to late blight is up 40 percent even with chemical control.

Figure 1.
Late Blight Affected Potato Crop

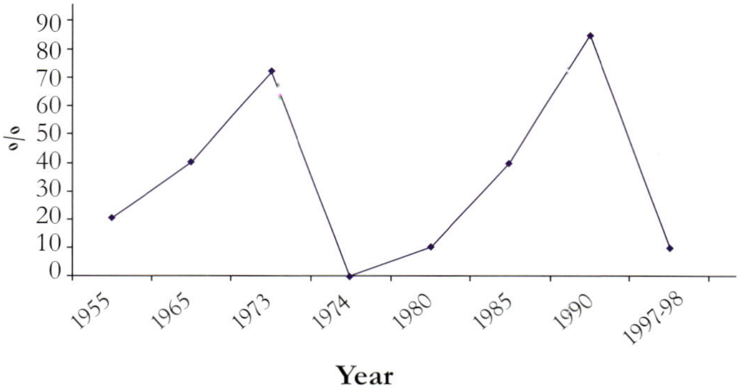

Figure 2.
Yield Loss due to Late Blight

TABLE 1.
Yield Loss due to Late Blight in Hills of Himachal Pradesh, India

Year	Losses (%)
1955	21
1965	40
1973	72
1974*	0.1
1980	10
1985	40
1990	85
1997-98**	10

Source: Potato Late Blight In India, CPRI (IARI) Technical Bulletin No.27 (Revised)
* Loss reduced due to introduction of *Kufri jyoti* (late blight resistant)
** Loss reduced due to introduction of *Kufri giriraj* (late blight resistant)
During 1980-90, losses were increased due to resistance break down

TABLE 2.
Variety Pattern and Yield losses due to Late Blight in Indian Plains

Year	Variety	% yield loss
1960-65	Susceptible	20
1975-80	Susceptible (KCM, KS)	40
	Resistant (KJ)	0
1985-90	Susceptible (KBR,KCM, KS)	65
	Resistant (KJ,KBD)	5-10
1996-97	Susceptible	80
	Resistant (KBD, *K sutlej, K anand*)	10-15

KCM= *Kufri chandramukhi*; KS=*Kufri sindhur*; KJ=*Kufri jyoti*; KBR=*Kufri bahar*, KBD=*Kufri bahar*; KBD= *Kufri badshah* .
Source: Potato Late Blight In India,CPRI (IARI) Technical Bulletin No.27 (Revised)

TABLE 3.
Impact of Late Blight on Potato Yield in India

Region	Varieties	Crop losses (%)	
Punjab	Susceptible	Sprayed	54.8
		Unsprayed	82.8
	Resistant	Sprayed	16.1
		Unsprayed	50.7
Western UP	Susceptible	Sprayed	–
		Unsprayed	50-60
	Resistant	Sprayed	–
		Unsprayed	25-35
West Bengal	Susceptible	Sprayed	40.0
		Unsprayed	80.0
	Resistant	Sprayed	5.0
		Unsprayed	10.0
Assam	Susceptible	Sprayed	10.0-12.0
		Unsprayed	20.0-25.0
	Resistant	Sprayed	1.0
		Unsprayed	5.0
Madhya Pradesh	Susceptible	Sprayed	50.0-60.0
		Unsprayed	70.0-80.0
	Resistant	Sprayed	35.0-40.0
		Unsprayed	65.0-70.0

Source: Late blight disease profile in sub-tropical plains (1998–99) (Singh and Shekhawat, 1999).

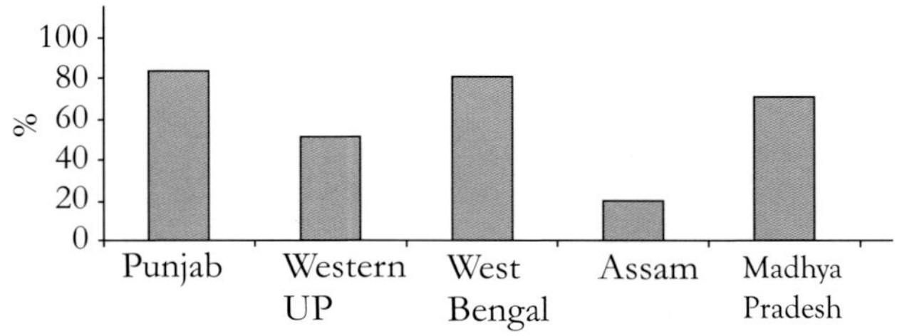

Figure 3.
Yield Loss Due to Late Blight in Major Potato Growing States in India

Fungicide requirements for controlling late blight in the hills where the crop is grown under rain fed conditions are substantial. Due to many applications of chemicals in frequent intervals, studies have found significant chemical residues on potato tubers. Of late, there has been an upsurge in awareness about environmental pollution and its impact on humans. Consequently, the scientific community must devise ways to reduce fungicides, which poses a serious threat to humans and the environment. As potato is one of the major food crops consumed by the majority of Indians, chemical residue free potatoes are highly demanded.

Disease Management

Fungicides alone will not help unless specific schedules are developed for all important potato growing regions keeping in mind the level of resistance in a given cultivar. Cultivar – specific fungicide schedules for both susceptible and resistant varieties have been developed for most of the popular cultivars in various regions, taking into account the time of appearance of the disease. In the sub-tropical plains the blight favorable period is generally limited to 1-2 weeks. Requirements of fungicides for controlling late blight in the hills where the crop is grown under rain fed conditions are high. Susceptible cultivars require a minimum of one spray of mancozeb. Better control is obtained with two sprays each of metalaxyl-based fungicides + 3 sprays of mangozeb. It was observed during our farmer survey that farmers also applied Ridomil to control the disease and spent more than Rs.1000 per hectare.

Cultural methods mainly aim at eliminating/reducing the initial inoculum load in seed tubers by cutting the infected haulms to avoid seed infection, harvesting tubers after proper skin care, avoiding collection of seeds from infected tubers, and sorting out of the diseased tubers before placing the seed tubers in cold/country stores and again before planting. In the hills, collection and destroying of infected tubers and vines, removing ground creepers or volunteer plants which harbor the pathogen, and formation of high ridges to cover the tubers thoroughly with soil, help prevent the fungus from the mother tuber infecting daughter tubers. In the plains, the disease normally appears with the onset of winter rains. Therefore, restriction of irrigation during that period and planning of alternative beds of susceptible and resistant varieties are the suggested cultural practices. Many of the farmers though aware of these practices, follow few of them due to many constraints. Most of the farmers are choosy in seed potato. Further, most of the farmers follow the practice of destroying infected plants. Farmers have a strong belief that applications of fungicides are needed to control late blight disease effectively.

Data Collection and Analysis

A farm survey was conducted in Himachal Pradesh covering 30 potato growers and 4 scientists in the Central Potato Research Institute, Shimla. Data pertaining to crop losses, cropping practices to manage the targeted problems, and crop yield and costs were gathered from farmers, experimental data, and opinions of agro-biological scientists. We asked about yield and costs with and without transgenic technologies, pesticide use, and time and cost of research. We also collected secondary data on price elasticities of demand and supply, output quantities and prices, trade, and pesticide use. Budgets were constructed and economic surplus and benefit cost analysis completed as described in Chapter 1.

Results

Economics of Production – Transgenic vs. Non-transgenic

Adoption of GM technology depends on the relative prices of conventional and GM seeds, and the cost of pesticides, labor, capital and other relevant inputs. For producers to have an incentive to adopt GM technology , production costs will have to decrease or remain static, or yields must increase. In addition, farmer diversity in terms of management ability, agronomic factors, and/or geographic location may determine the extent of economic gain from GM crops, as producers differ in their tolerance for risk {Kalaitzandonskes, 2003, Fulton, M., & Keyowski, L. (1999)}. Our survey found that Indian farmers on average used 2585 kg of seed per hectare and the cost of seed alone accounted for more than 50 percent of the total cost of cultivation. Scientists who were interviewed believe there would not be any reduction in the amount of seed used in the case of transgenic potato, although farmers may pay a higher price for seed. A technology cost for the GM product is typically passed on to farmers through the seed price, in this case estimated to be 20 percent higher. Estimates of costs and returns reveal that adopting transgenic LBR potato would reduce pesticide application costs by Rs. 1100 and result in a lower cost of labour and higher yield. Farmers would benefit from adopting LBR potato, as their net income would double as compared to using existing varieties such as *Kufri jyoti*, *Kufri giriraj* and *Kufri bahar.* There would be a reduction in labour use of 11 percent due to reduced application of fungicides. Yield of LBR potato would be 25 percent higher. Farmers would incur Rs. 73246 per hectare in cost of cultivation of LBR potato, as compared to Rs. 68893 for existing varieties due to higher seed costs for the GM potato, but total returns would be Rs. 190,000 per hectare in the case of LBR potato, as compared to Rs.127,000 for existing varieties (Table 4 and Figure 4).

TABLE 4.
Cost and Returns of Transgenic (LBR Potato) and Non-Transgenic(Rs./ha) in India

Cost	Non-Transgenic	Transgenic	Difference (%)
Seed	36265 (52.64)	43517.5 (59.40)	(+)20.00
Labour	17100 (24.81)	15300 (20.89)	(-)10.53
Bullock Labour	3532 (5.13)	3532 (4.81)	–
FYM	1575 (2.29)	1575 (2.14)	
Calcium Ammonium Nitrate	3775 (5.48)	3775 (5.14)	
Single Super Phosphate	3805 (5.51)	3805 (5.18)	
Potash	1375 (2.00)	1375 (1.88)	
Pesticides			
Insecticides (Carbofuran and others)	366 (0.53)	366 (0.50)	–
Fungicides (Ridomil)	1100.25 (1.60)	–	–
Total Cost	68893.25 (100.00)	73245.5 (100.00)	(+) 6.32
Return			
Yield (tonnes)	13.79	17.24	(+)25.00
Price (000 Rs/tonnes)	9.18	11.04	(+)20.00
Gross Income (000 Rs/ha)	126.58	190.30	(+)50.34
Net Income (000 Rs/ha)	57.69	117.04	(+)102.88
Cost Benefit ratio	1.84	2.60	

* Predominant varieties are - *Kufri jyotii, Kufri giriraj, Kufri bahar*
** 20% increase in product price was assumed based on the low level of pesticide application, as consumers are assumed to pay more for produce free of pesticide residues.

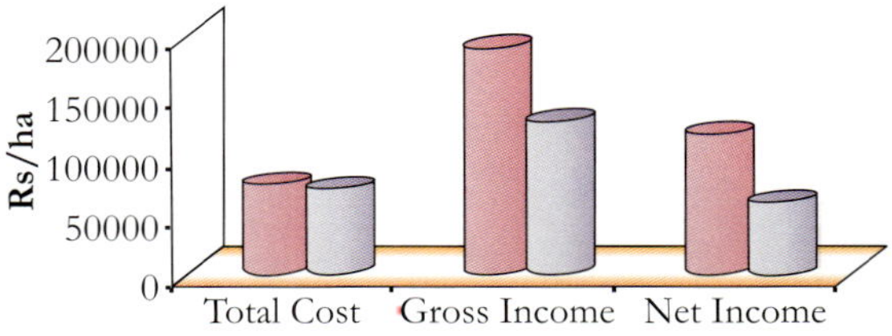

Figure 4.
Costs and Returns of Transgenic (LBR) and Non - Transgenic Potato

Return on Investment

According to scientists, about 9 years are needed for technology development and the probability of success would be 90 percent. Once the technology is developed and commercialized after meeting the regulatory requirements, the spread would be very fast and it is expected that about 45 percent of the area under potato would be covered with transgenic potato due to its yield advantage and cost reduction. Scientists were of the opinion that cost and time required to complete the research and meet the regulatory hurdles are higher and more expensive than technology development. Costs of meeting the regulatory requirements such as conducting field and multi-location trials and the cost of food safety assessment and commercialization are high (14.25 million rupees (US $ 0.36 million) Using the economic surplus estimates as benefits and the technology and regulatory costs as the costs, NPVs and IRRs were projected (Table 5). The projected NPV over 17 years ranges from Rs. 19398 million (485 million US $) to Rs. 70745 million (1769 million US $) while the IRR ranges from 123 percent to 164 percent assuming open trade with the closed economy assumption, the projected NPV varies from Rs. 31224 million (781 million US $) to Rs. 190290 million (4757 million US $) and the IRR ranges from 137 percent to 187 percent. These results indicate that investment in developing and commercializing LBR potato would bring high economic and social benefits.

TABLE 5.
Estimated NPV and IRR for LBR Potato

Scenarios	**Open Economy Model**		**Closed Economy Model**	
	NVP (Rs in million)	**IRR (Percent)**	**NPV (Rs. in million)**	**IRR (percent)**
Relatively Inelastic Supply (**e=0.4)**	70745 (1769)	164	190290 (4757)	187
Unit elastic **(e=1)**	24533 (613)	130	39482 (987)	145
Relatively elastic **(e=1.2)**	19398 (485)	123	31224 (781)	137

(Figures in parentheses are in million US dollars)

Conclusion

Reported yield losses due to LBR even for the resistant varieties are high, in some states up to 40 percent, even with chemical control. The findings of this study are that farmers would benefit significantly from adopting LBR potato, as their net income would be twice that of existing varieties such as *Kufri jyoti, Kufri giriraj* and *Kufri bahar*. Yield of transgenic LBR potato would be 25 percent higher than existing varieties. Estimates of NPVs and IRRs are high, indicating that investment in developing and commercializing transgenic LBR potato would bring high economic and social benefits. There should also be health and environmental benefits due to reduced fungicide use.

Reference

Flannery, M.L., Thorne, F.S., Kelly, P.W. and Mullins, E (2004). An Economic Cost-Benefit Analysis of GM Crop Cultivation: An Irish Case Study. *AgbioForum,* 7(4). Pp149-157. www.agbioforum.org

Fulton, M., & Keyowski, L. (1999). The Producer Benefits of Herbicide-resistant Canola. *AgBioForum*, 2(2), 85-93. Available on the World Wide Web: www.agbioforum.org.

Kalaitzandonakes, N.G. (2003). The Economic and Environmental Impacts of Agbiotech: An Introduction. In N.G. Kalaitzandonakes (Ed.), *The Economic and Environmental Impacts of Agbiotech* (pp.1-18). New York: Kluwer Academic.

Singh, B.P.and Shekhawat, G.S.(1999). Potato Late Blight in India. *Technical Bulletin No.27*. Central Potato Research Institute, Shimla, India. PP.85

Chapter **5**

Potential Socioeconomic Impacts of Bt Eggplant in India

V.V. Krishna and M. Qaim

Introduction

Eggplant is one of the most prominent vegetable crops in India – grown mostly by smallholder farmers on a total of 1.3 million acres. Although India ranks second after China in worldwide eggplant production, crop productivity is relatively low. Under the climatic conditions in India, eggplant is infested by a number of insect pests, the most destructive of which is the eggplant shoot and fruit borer (ESFB, *Leucinodes orbonalis* Guen.). Despite heavy insecticide applications, significant yield losses occur on a regular basis. Bt eggplant, containing the Cry1Ac gene, which provides resistance to the ESFB, has been developed by the Maharashtra Hybrid Seed Company (MAHYCO). Several Bt hybrids have been tested in the field and are likely to be commercialized in the near future. Besides commercializing Bt eggplant hybrids itself, MAHYCO has shared its technology and know-how free of charge with the public sector, which is now backcrossing the Bt gene into open-pollinated varieties (OPVs) of eggplant. This is sponsored by the United States Agency for International Development (USAID) through its Agricultural Biotechnology Support Project (ABSPII). The rationale behind this private-public technology transfer is that relatively better-off farmers would adopt Bt eggplant hybrids sold by MAHYCO, while resource-poor farmers would use cheaper OPVs provided by the public sector.

This chapter reports on an analysis of potential socioeconomic impacts of Bt eggplant technology in general and the private-public technology transfer in particular. To structure the research, the overall objective was subdivided into six research questions as follows:

1. What are likely productivity and pesticide reduction effects of Bt eggplant technology in India?
2. What is the potential demand for Bt technology in eggplant production, and how do likely adoption patterns look like?
3. What are the prospective environmental benefits due to reduction in pesticide use?
4. What are the probable impacts of Bt eggplant on the nutrition status of consumers?
5. What are consumer attitudes towards transgenic vegetables?
6. What are the potential aggregate welfare and distribution effects of Bt eggplant in India?

Data collection

Data for the empirical research have been obtained from five major sources:

- Field trials with Bt eggplant carried out by MAHYCO,
- Interviews with researchers, agronomists, and other experts,
- A survey of eggplant farmers,
- A survey of eggplant-consuming households,
- Secondary data sources and statistics,

These data sources are briefly described in the following.

Trial data

Since 2004, MAHYCO has been testing 8 different Bt eggplant hybrids in several states of India. These multi-location field trials were managed by company researchers to test the agronomic and biosafety performance of the new technology. In each location, a Bt hybrid was grown next to an isogenic non-Bt hybrid and other conventional checks, including both popular eggplant hybrids and OPVs. MAHYCO researchers kindly shared with us their data from nine locations in the 2004-05 season and from six locations in the 2005-06 season.

Interviews with researchers, agronomists, and other experts

We had interviews and comprehensive discussions with MAHYO researchers, scientists from different institutes of the Indian Council for Agricultural Research (ICAR), and several representatives of other Indian and international institutions (including the World Vegetable Center, AVRDC). These discussions helped to better understand pest problems in eggplant production and details of the transgenic technology. Moreover, the meetings were very useful to comprehend the Indian vegetable sector in general, including seed market and other institutional aspects. Finally, the sampling frameworks for the farmer and consumer surveys were also discussed with different experts.

Farm survey

A comprehensive questionnaire was developed for the farm survey, which concentrated on input-output relationships in eggplant production and information needed to forecast productivity and future technology adoption. Furthermore, questions on the farmers' willingness to pay (WTP) for Bt hybrids and Bt OPVs and household characteristics were included. In total, 360 eggplant farmers were interviewed in three leading eggplant-producing states of India – Andhra Pradesh, West Bengal, and Karnataka. Data collection was carried out during the period between February and May 2005. Based on expert assessments, the sample can be considered representative of the major eggplant-growing regions of India.

Consumer survey

Realizing that consumer acceptance issues might become a crucial factor for the success of Bt eggplant technology in India, we felt that a larger consumer survey would be helpful, allowing us to carry out comprehensive and representative statistical data analysis. Since the ABSPII grant would not have allowed us to implement a large survey of consumers in different parts of

India, we were glad to secure some limited matching funds from AVRDC. Since AVRDC was also interested in consumer acceptance issues, we agreed to increase the sample size and scope of the consumer survey that we had initially planned. Yet the responsibility for survey planning and implementation remained with us. Data were collected between February and April 2006 from five urban locations in India, viz. New Delhi, Bangalore, Kolkata (Calcutta), Kolar, and Barddhaman (Burdwan). The first three locations are among the largest cities of India and administrated by municipal corporations, whereas Kolar (Karnataka) and Barddhaman (West Bengal) are two district headquarters in close proximity to important vegetable production regions. These locations were selected purposely, as they provide a good representation of urban India. Overall, 645 households from 61 corporation wards were interviewed. The respondents were usually the persons in charge of household vegetable purchases.

Secondary sources and statistics

A careful review of secondary sources was undertaken, and the data were used for our simulations and projections. In particular, data from the National Horticulture Board in India and the Food and Agriculture Organization of the United Nations were used. Moreover, previous literature on the national vegetable sector as well as on Bt cotton experiences in India and other countries was used to make realistic assumptions in the ex ante analysis.

Results

Research results are documented in the following three papers (Krishna and Qaim, 2007a, 2007b, 2008) and are summarized here, following the order of the research questions posed above. For further details, reference is made to the referenced paper.

Productivity and pesticide reduction effects of Bt eggplant technology in India

Table 1 summarizes the trial performance of Bt hybrids over two years. The technology allows significant insecticide reductions: on average, amounts of insecticides used against ESFB were reduced by 80%, which translates into a 42% reduction in total insecticide quantities. At the same time, there is a large positive yield effect, indicating that chemical insecticides are only of limited effectiveness in controlling ESFB losses. Yields of Bt hybrids were double those of non-Bt counterparts; the yield advantage with respect to other popular hybrids and OPVs was even more pronounced.

TABLE 1:
Summary of field trial results with Bt eggplant hybrids

	Reduction in insecticide use (%)		Increase in uninfected fruit yield (%) over		
	Against ESFB	Against all insect pests	Non-Bt counterparts	Popular hybrids	Popular OPVs
2004-05 (*n* = 9)	80	44	117	120	179
2005-06 (*n* = 6)	79	40	76	110	147
Average	80	42	100	116	166

Source: MAHYCO (unpublished data).

While the results of these researcher-managed trials indicate that Bt eggplant technology could lead to important agronomic advantages in the Indian vegetable sector, they might not be replicable exactly under typical farmer conditions. Crop enterprise budgets per acre of eggplant are shown in Table 2, separately for the production regions in the Center/South (mainly comprising the states of Karnataka, Andhra Pradesh, Maharashtra, Tamil Nadu, Gujarat, Madhya Pradesh) and East (West Bengal, Orissa, Bihar, and Assam). The "without Bt" columns show the situation as currently observed based on farm survey data. The "with Bt" columns show projections which are based on the field trial results and appropriate adjustments to account for practical farmer conditions. These adjustments were made jointly with local vegetable experts. It is assumed that insecticide use against ESFB would be reduced by 75%, which results in a total reduction in insect pest management cost of 35% and 48% in Center/South and East, respectively. Increases in effective yields through Bt eggplant technology are assumed to be significant, but much lower than those observefarmers' fields, yield advantages of Bt eggplant hybrids over conventional hybrids are expected to be around 40%, whereas the advantage of Bt hybrids over conventional OPVs could be around 60% (including hybrid vigor and the Bt gene effect). The calculations in Table 2 take into account the current proportion of hybrid use in the two regions (3).

TABLE 2:
Average eggplant enterprise budgets with and without Bt hybrids

	Center/South		East	
	Without Bt	With Bt	Without Bt	With Bt
Seed cost (Rs/acre)	638	4,642	48	4,642
Insecticide cost (Rs/acre)	1,972	1,282	6,776	3,543
Labor cost for insecticide sprays (Rs/acre)	186	121	499	261
Harvesting/marketing cost (Rs/acre)	4,049	5,993	1,309	2,094
Other cost (Rs/acre)	11,052	11,052	14,462	14,462
Total variable cost (Rs/acre)	17,897	23,090	23,094	25,002
Marketable yield (quintals/acres)	106	157	71	114
Per-unit production cost (Rs/quintal)	169	147	325	220
Gross revenue (Rs/acre)	44,670	66,162	32,907	52,651
Gross margin (Rs/acre)	26,773	43,072	9,813	27,649

Source: (3).

MAHYCO has not yet fixed the price at which Bt hybrid seeds will be sold in future. Using contingent valuation techniques, it was estimated that farmers' mean WTP for Bt eggplant hybrids is Rs. 4,642/ acre (1). This is about five times the price of conventional hybrids, which are sold at around Rs. 900/acre, and a multiple of the cost of OPV seeds. Nonetheless, given the large agronomic advantages and comparing with the price difference between Bt and conventional cotton seeds in India, the magnitude appears reasonable, so that we assume that Bt eggplant hybrids would be priced at mean WTP. The resulting effects of Bt technology on cost of production are shown in Table 2.

While the cost per acre of eggplant increases slightly, the cost of production per unit of output is reduced by 13% in the Center/South and 32% in the East. Assuming constant output prices, gross margins are expected to increase by Rs. 16,299/acre (US $361) and Rs. 19,744/acre (US $437) in the two regions (3).

Potential demand for Bt technology in eggplant production and likely adoption pattern

To elicit eggplant farmers' WTP for Bt hybrid seeds, the contingent valuation method was employed. A dichotomous choice approach was used. Bt eggplant technology was explained in detail to all farmers, before they were asked whether they would be willing to use Bt hybrids at a certain price level. Accordingly, mean WTP for Bt hybrids was estimated at Rs. 4,642/acre (1). At that price level, we project that around 50% of all eggplant farmers would adopt Bt hybrids. Farmers' price responsiveness is shown in Figure 1.

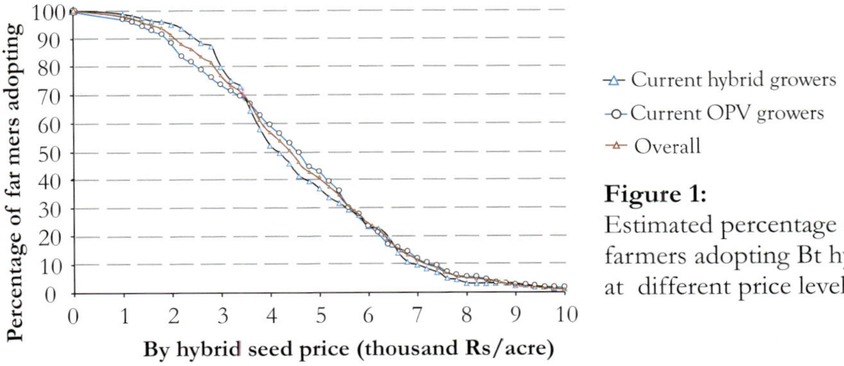

Figure 1:
Estimated percentage of farmers adopting Bt hybrids at different price levels

Once Bt OPVs enter the Indian seed market after a small time lag, the adoption process will probably become more complex. Farmers will have three options: (i) adopting Bt hybrids, (ii) adopting Bt OPVs, and (iii) non-adoption of the technology. To elicit farmers' preferences in this changed scenario, a simple choice experiment was implemented (1). Figure 2 shows estimated technology adoption rates with and without the existence of Bt OPVs, assuming that both types of seeds are priced at there mean WTP. As can be seen, some of the Bt hybrid adopters would switch to Bt OPVs, once these become available.

Nonetheless, Bt hybrids would still be used by 37% and 24% of farmers in the Center/South and East, respectively. Total adoption rates for Bt technology (incorporated in hybrids and OPVs) are expected to increase to 78% in the Center/South and 93% in the East.

Figure 2:
Estimated adoption rates at mean WTP for Bt hybrids and Bt OPVs

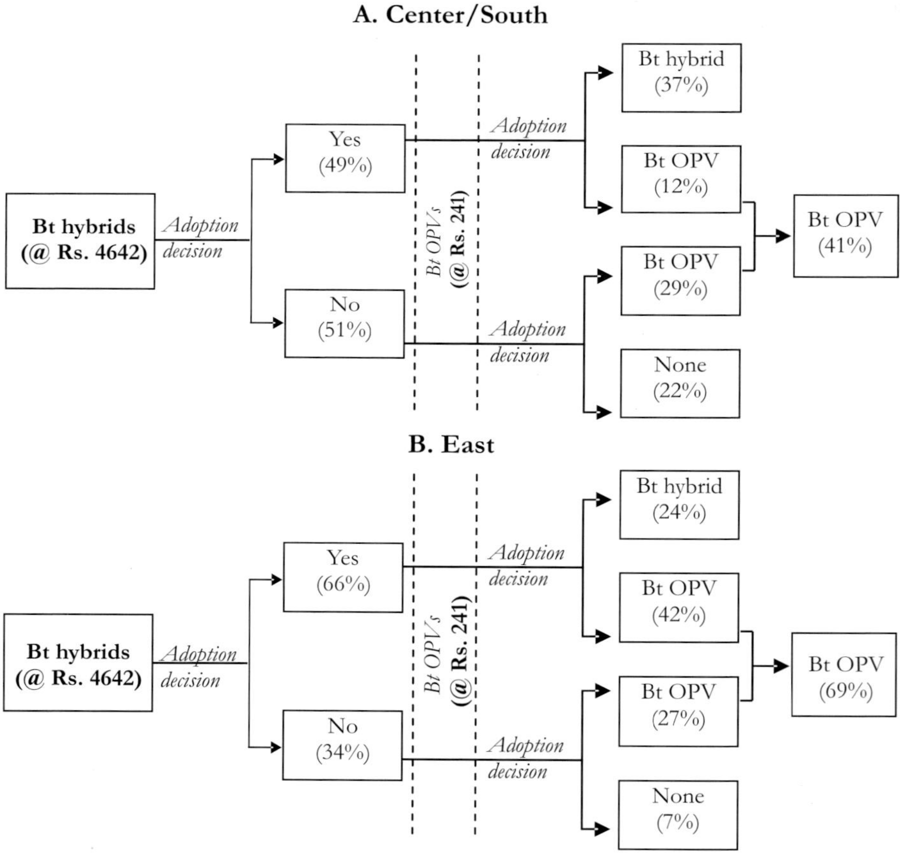

Prospective environmental benefits due to reduction in pesticide use

Compared to most developed countries, pesticide use in Indian agriculture is relatively low. This is different, however, for certain crops and regions. Fruit and vegetable crops, in particular, are sprayed quite heavily in India: while they only cover 3% of the gross cropped area, they receive 13% of total pesticides used. A typical farmer applies 30 insecticide sprays during a single eggplant crop of 180 days. Repeated application of pesticides on eggplant often results in buildup of residues. Different studies indicate high levels of insecticide residues in eggplant, with detrimental health effects for consumers. Often, residues of pesticides that are banned in many other countries, such as monocrotophos, HCH, carbofuran, and endosulfan, were observed. The use of highly toxic products is confirmed by our farm survey data. The most frequently used insecticide include organochlorins, organophosphates, and carbamates, which are known for their high mammalian toxicity. The most popular insecticides used among sample farmers are endosulfan and monocrotophos; both fall into toxicity category I of the World Health Organization's classification. Apart from serious negative environmental externalities and health hazards for consumers, manual pesticide applications are affecting the health of eggplant farmers themselves.

The majority of eggplant farmers are aware of potential health hazards associated with pesticide applications, and around one-fourth of them have suffered personally from acute pesticide poisonings during the 12 months prior to data collection. The cost of illness caused by pesticide poisonings was calculated to study the potential impacts of Bt technology on farmers' health. Using an econometric model, health cost savings resulting from insecticide use reductions through Bt technology were predicted. The results are shown in Figure 3. With the expected insecticide reductions through Bt eggplant technology in the Center/South (35%) and East (48%), health cost savings would be around Rs. 50/acre and Rs. 470/acre, respectively. The big regional differences are due to the higher initial number of insecticide applications and pesticide poisonings in the East

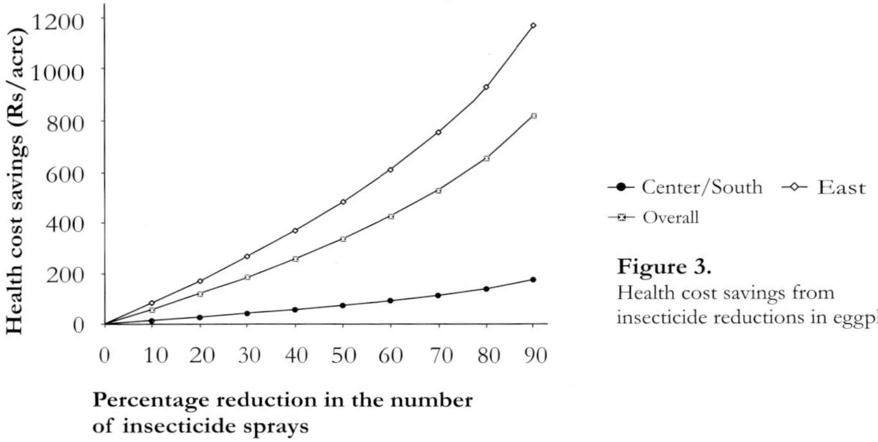

Figure 3.
Health cost savings from insecticide reductions in eggplant

Aggregate health cost savings through Bt were calculated by multiplying the per acre savings with projected technology adoption rates. Average annual health cost savings from Bt hybrid adoption would be Rs. 135 million (US $3 million). When Bt OPVs are introduced additionally, aggregate health cost savings would further increase to Rs. 184 million (US $4 million) per year (3). These are large benefits, which are often neglected in economic analyses. It should be stressed, however, that these savings are only a fraction of the overall potential environmental and health benefits of Bt eggplant technology.

Probable impacts of Bt eggplant on the nutrition status of consumers

The widespread adoption of Bt technology will lead to a decrease in market prices for eggplants. Lower prices, in turn, lead to higher consumption of eggplants with positive nutrition effects. We project that in the longer run, at maximum technology adoption rates market prices of eggplant would decrease by about 15% . At the observed price responsiveness of consumers, this would lead to a 4% increase in eggplant quantities consumed on average. The effect is likely to be more pronounced among the poor, since eggplant in India is an important vegetable also in low-income consuming households, whose price responsiveness is larger than that of high-income households. Hence the technology will have a pro-poor effect among Indian food consumers.

Consumer attitudes towards transgenic vegetables

Consumer concerns and risk perceptions have emerged as important driving forces of biotechnology policies in developed countries. For developing countries, political pressures and uncertainty regarding consumer acceptance were important reasons for the initial slow spread of transgenic crops. The pest resistance imparted through genetic engineering could help farmers reduce their insecticide use, also resulting in lower levels of toxic residues. But will consumers appreciate this reduction in pesticide residues, or will their attitude be influenced primarily by the fact that the vegetables are transgenic, so that they might potentially be associated with new types of risks, at least in consumers' perceptions? This question was addressed, building on the consumer survey data.

Most households are well aware of residue-related health risks, and on average consumers are willing to pay 57% more for residue-free vegetables. The majority of households also have a positive attitude towards the introduction of Bt technology. Table 3 shows that 68% generally support the introduction of Bt vegetables, while 17% mildly or strongly oppose the new technology. With few exceptions, the variation in attitudes across the different locations is relatively low. The rather positive perception of the technology in India is consistent with other studies, which showed that the attitude towards agricultural biotechnology in developing countries is generally less negative than in developed countries, and often even positive. One plausible explanation is that people in developing countries are generally poorer and sometimes food-insecure, so that they are more dependent on productivity-increasing agricultural technologies than better-off consumers in developed countries.

TABLE 3:
Consumer attitudes towards the introduction of Bt vegetables

Attitude towards Bt vegetables	% of households in					
	New Delhi	Bangalore	Kolar	Kolkata	Barddhaman	Overall
Strongly supporting	13.50	11.46	18.75	49.70	10.00	22.48
Mildly supporting	47.85	57.96	51.25	26.06	53.75	45.89
Indifferent	13.50	8.28	7.50	12.12	13.75	11.16
Mildly opposing	11.66	9.55	17.50	3.64	18.75	10.70
Strongly opposing	9.20	5.73	2.50	7.27	2.50	6.20
Can't tell	4.29	7.01	2.50	1.21	1.25	3.57
Overall	100.00	100.00	100.00	100.00	100.00	100.00

Figure 4 shows that almost 60% of respondents in our sample would purchase Bt vegetables at current market prices of conventional vegetables. More than 80% would purchase Bt vegetables at a price discount of 10% (2). Unsurprisingly, respondents who generally oppose the technology would require larger price discounts in order to purchase Bt vegetables. Still, around 30% of them would be willing to purchase also without any discounts, indicating that the opposition is not very pronounced. This bodes well for the future introduction of Bt eggplant technology in India.

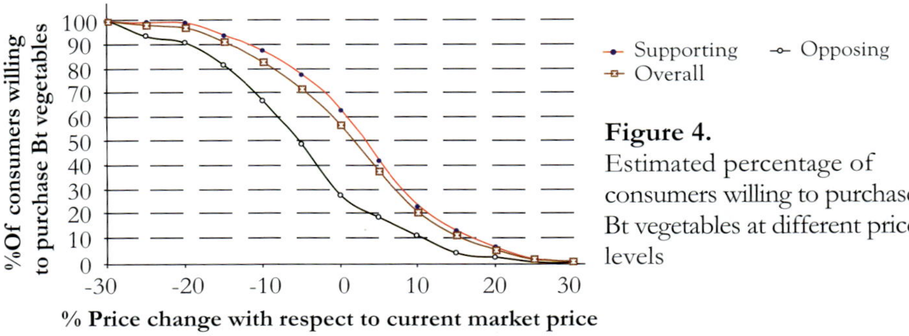

Figure 4.
Estimated percentage of consumers willing to purchase Bt vegetables at different price levels

Strikingly, however, the more households are willing to pay for residue-free products the less they are willing to accept Bt vegetables, that is, the WTP for residue-free vegetables is negatively correlated with the WTP for Bt vegetables. Obviously, risk aversion as such is a main driving force for consumer decisions, and a reduction in one risk cannot easily compensate for an increase in another. More generally speaking, consumers who are concerned about the potential risks of a new technology may act irrationally by undervaluing tangible benefits (3).

Potential aggregate welfare and distribution effects of Bt eggplant in India

At first, we simulate the economic surplus effects of Bt eggplant in a scenario where the technology is only incorporated in hybrids sold by MAHYCO, that is, Bt OPVs are not available. The results are shown under scenario I in Table 4. The total surplus generated by the technology amounts to an annual average of Rs. 4.9 billion (US $108 million). This is a large benefit in absolute terms for a vegetable with an aggregate area coverage which is much smaller than that of major food or fiber crops. The welfare gains would further increase if environmental and health benefits are added to the economic surplus effects.

In terms of surplus distribution by economic agents, consumers turn out to be the main beneficiaries. Because of a closed-economy assumption and relatively inelastic consumer demand for eggplant, Bt technology would lead to an average price drop of around 15% at maximum adoption rates. However, eggplant producers also benefit, in spite of the negative price effect. Their annual gain amounts to Rs. 0.7 billion (US $15 million). Of course, farmers' share of the surplus gains could rise if eggplant exports from India were promoted. MAHYCO as the innovating company captures around one-third of the total surplus in the form of gross technology revenues.

TABLE 4.
Simulated gains in economic surplus and surplus distribution (annuities in m Rs)

		Surplus distribution				
		By region		By economic agent		
	Total	Center/South	East	Consumers	Producers	Company
Scenario I	4,896	888	4,008	2,716	679	1,501
		(0.18)	(0.82)	(0.55)	(0.14)	(0.31)
Scenario II	1,008	888	120	416	104	487
		(0.88)	(0.12)	(0.41)	(0.10)	(0.48)
Scenario III	5,676	1,112	4,564	3,886	972	819
		(0.20)	(0.80)	(0.68)	(0.17)	(0.14)

Notes: US $1 = Rs. 45.2 (official-exchange rate in October 2006). Figures in parentheses indicate the share of total economic surplus.

To analyze the implications of different institutional aspects, two additional scenarios were simulated. Like scenario I, scenario II assumes that only Bt eggplant hybrids are available. But particular account is taken of the fact that vegetable seed markets are poorly developed in the East. Underdeveloped seed markets are partly responsible for the low current use of eggplant hybrids in that region. Yet another reason is that certain vegetable-producing pockets of Eastern India are particularly affected by soil-borne fungal pathogens, to which many of the available eggplant hybrids are more susceptible than OPVs. For illustrative purposes, we assume a maximum adoption rate of only 2% for the East, while adoption in the Center/South remains unchanged. As expected, the aggregate surplus gains are much smaller, only reaching about one-fifth of those in scenario I. This clearly shows that Eastern states should receive high priority in technology development and delivery strategies. Apart from putting extra effort in the establishment of local seed market infrastructure, this also involves the deliberate incorporation of Bt technology into hybrids suitable for the particular agroecological conditions of Eastern India.

Scenario III analyzes the implications of the private-public technology transfer. Given that Bt OPVs are still at a somewhat earlier stage of development, it is expected that they will be commercialized with a small delay. As can be seen in Table 4, with Rs. 5.7 billion (US $126 million) the total annual surplus gain in scenario III is bigger than in the other two scenarios. As in scenario I, the largest benefit accrues in the East, and in terms of economic agents, eggplant consumers are the main beneficiaries. However, also the producer benefit is increased both in absolute and relative terms. As intended by the private-public sector agreement, especially resource-poor producers will profit from the introduction of Bt OPVs. Thus, apart from leading to higher overall welfare gains, MAHYCO's technology transfer also further improves the equity effects of Bt eggplant technology.

Yement cet, the agreomes at a cost for the company. Gross technology revenues still occur, but they are much lower than they would be without the transfer. Against this background, the rationale to share the technology with the public sector is not immediately apparent. Yet the scenario calculations do not capture all direct and indirect implications of the agreement. Corporate social responsibility can pay off in multiple ways, especially in an environment like India, where the biotechnology debate is highly politicized and public distrust towards large private companies is widespread. Apart from general image improvements, the agreement could facilitate regulatory approval procedures and lessen the probability of public calls for government price interventions, as recently observed in Bt cotton seed markets. These considerations suggest that private-public technology transfers can well be beneficial for all parties involved, if appropriately designed and managed.

Reference

Krishna, V.V. and M. Qaim. "Estimating the Adoption of Bt Eggplant in India: Who Benefits from Public-Private Partnership?" *Food Policy,* 2007 (in press).

Krishna, V.V. and M. Qaim. "Consumer Attitudes Towards GM Food and Pesticide Residues in India" *Review of Agricultural Economics,* 2008 (in press).

Krishna, V.V. and M. Qaim. "Potential Impacts of Bt Eggplant on Economic Surplus and Farmers' Health in India," Department of Agricultural Economics and Social Sciences, University of Hohenheim, Stuttgart, Germany, 2007.

Chapter **6**

Salinity and Drought Tolerant Rice in Bangladesh

S.M. Fakhrul Islam and G.W. Norton

Introduction

Soil salinity is a major threat to irrigated agriculture in Bangladesh, with around 30% of the net cultivated land in the coastal area. Of the 2.85 million ha of coastal and off-shore lands, about 0.83 million ha of arable land are affected by varying degrees of soil salinity, which constitutes about 52.8 percent of the net cultivable area in 64 upazillas of 13 districts (Karim, et al., 1990). Coastal areas of Bangladesh vary widely in terms of soil salinity and depth of groundwater, as well as salinity of surface water (MPO, 1986). In saline soils, crop growth is hampered by salt accumulation in the crop root zone. If the upward salt movement caused by evaporation exceeds the downward gravitational movement, salt will accumulate in the root zone. In the coastal region of Bangladesh, a considerable amount of salt accumulates on the soil surface by evaporation, particularly in the presence of a shallow saline water table during the dry period from February to May when the land remains fallow. If the fallow period is long, severe salinity may develop.

During the past few decades, modern agricultural technologies based on high-yielding seeds, improved fertilizer and water management, and improved insect and disease control measures have helped to increase crop production in most of Bangladesh. But the coastal regions of Bangladesh have remained deprived of this technological advancement mostly because of soil salinity, poor drainage and lack of good quality irrigation water (Ghosh et al., 1987; Karim et al., 1990). Consequently cropping intensity and production levels are much lower in this region than in other parts of the country (BBS, 1996). This problem has received little attention by researchers and it is now imperative to more fully exploit the potential of these lands for meeting increased food demand and for improving the socio-economic conditions of resource-poor farmers. As the population continues to grow, demands for freshwater for non-agricultural uses will rise, which will lead to a reduction of agriculture's share of freshwater (Bhuiyan, 1997). Increasingly, agriculture will face the challenge of having to use water of poorer quality (Oster, 1994). Therefore, an intelligent management of coastal saline soils and brackish groundwater for producing more food crops will become more important in the future.

Economic impact information on genetically modified (GM) crops is available for the principal GM crop-growing countries (Commission of the European Communities, 2000, Dowley, et al., 2001; Falck-Zepeda, J.B., Traxler, G., and R.G. Nelson, 2000; Tolstrup, et al., 2003; Van Meijl and Tongeren, 2003; Demont, et al., 2004a and 2004b; Alston, J. M, Hyde, J., Marra, M.C., & Mitchell, P.D., 2002). For Bangladesh, such information is not yet available. The purpose of the present study is to provide an assessment of the potential economic impacts of transgenic salinity and drought resistant (SDR) rice in Bangladesh. The objective is to predict costs and/or benefits that a producer might experience if a transgenic SDR rice were to be cultivated in coastal area of Bangladesh, as well as to project the aggregate benefits to society. For producers, an economic incentive is an essential factor in deciding to adopt or reject a new technology (eg. GM crops). For this research, it is assumed that producers will base their adoption decision on the relative prices of conventional and GM seeds, chemical pesticides, labour, capital, and other relevant inputs and choose a system that will minimize these costs per ton of production. If producers are to be motivated to adopt a GM technology, production costs per ton will have to decrease (Kalaitzandonakes, 2003). In addition, farmer diversity, in terms of management ability, agronomic factors, and/or geographic location (Fulton & Keyowski, 1999), will determine the extent of economic gains from GM crops, as producers will differ in their tolerance for risk (Kalaitzandonakes, 2003). Consequently, it can be concluded that some producers will benefit from the new technology while others may not (Fulton & Keyowski, 1999). Partially offsetting the anticipated cost savings associated with the adoption of a GM technology, the producer will also bear some additional costs. A technology cost typically will be passed to farmers in the form of a seed premium (PG Economics, 2003) as demonstrated in the United States and Argentina (US General Accounting Office, 2000), Spain (Demont & Tollens, 2004b), and the United Kingdom (May, 2003). Some other associated costs such as harvest labor may also increase (Tolstrup et al., 2003).

METHODS

Data sources

Given the intensity of rice cultivation in the salinity-affected coastal region, Noakhali and Sathkhira districts were selected for field surveys for the present study. Data were collected from 60 farmers, 30 from each region. Data were also collected from 5 scientists and 8 industry experts from the Department of Agricultural Extension (DAE), the Bangladesh Agricultural Development Corporation (BADC), and seed companies. Data were collected from farmers on input costs such as seed, fertilizers, pesticides and labor. Data were collected on crop varieties, sources of seeds, crop management practices, input and output prices, crop losses due to salinity problem, and crop yields.

Data were collected from scientists on potential regions for GM SDR rice, expected yield changes if GM SDR rice were grown, changes in variable costs for GM SDR rice, time lags and costs for technology development and regulatory requirements.

From industry experts, data were collected on preferred varieties, sources of seed, extent of crop losses due to salinity and drought, expected extent and time path of adoption of GM SDR rice, time lags and costs for technology development and regulatory requirements. Secondary data on crop acreage and production were collected from district offices of DAE and various issues of Year Book of Agricultural Statistics of Bangladesh.

Economic surplus and benefit cost analysis

Cost and return information, secondary data on prices and quantities, and adoption information were combined in an economic surplus model to project total economic benefits and their distribution by region within the country and to consumers, producers, and seed sector (Alston, Norton, and Pardey 1995). A small open economy was assumed. The costs of the research and product development (including meeting regulatory hurdles) were obtained from scientists and included along with the benefits in a benefit cost analysis of the public investment. Excel spreadsheets were used for estimating the economic surplus models and net present value of benefits over 15 years.

RESULTS AND DISCUSSION

The extent of salinity area and trend in rice acreage and yield

Table 1 presents the distribution of and extent of different categories of soil salinity in the 13 costal and offshore districts of Bangladesh. Table 2 presents the potential cropping patterns for different land types and varied soil salinity levels in the 13 coastal districts.

Figure 1 present trend in area, production and yield of Aus rice cultivated in the salinity-affected areas of Noakhali, Greater Khulna (Sathkhira included), and Barisal districts. In these areas, Aus rice is cultivated during the summer season and suffers from salinity problems. It is observed that salinity rice acreage in Noakhali district was almost stable, around 120,000 ha, but its production and yield fluctuated because of salinity during 1990-91 to 2003-04. The yield of salinity affected Aus rice is found to be very low and it fluctuated from 1000 to 1500 kg/ha during the same period.

In Khulna district, acreage and production of salinity affected rice gradually decreased from 1990-91 to 2003-04. Its yield was also low and fluctuated around 1200 to 1600 kg/ha. In Barisal acreage of salinity affected Aus rice was almost stable, around 150,000 ha, but yield and production were found to have an increasing trend from 1998-99 to 2003-04. From 1990-91 to 1998-99, the yield of salinity affected Aus rice fluctuated around 1000 kg/ha. Beyond this period, yield increased and was about 1400 kg/ha in 2003-04.

TABLE 1.
Distribution of and extent of different categories of soil salinity in the costal and offshore regions of Bangladesh. ('000'Hectares)

District		Salinity categories				Total Area
		S1	S2	S3	S4	
Satkhira	Area	16.5	85.6	33.35	10.9	146.35
	%	11.3	58.5	22.8	7.4	100
Khulna	Area	3.9	92.54	13.8	9.8	120.04
	%	3.2	77.1	11.5	8.2	100.0
Bagerhat	Area	28.3	77.08	2.6	0	107.98
	%	26.2	71.4	2.4	0.0	100.0
Barguna	Area	96.39	7.2	0	0	103.55
	%	93.1	7.0	0.0	0.0	100.0
Patuakhal	Area	68.5	46.6	0	0	115.1
	%	59.5	40.5	0.0	0.0	100.0
Bhola	Area	9.52	30.81	0	0	40.33
	%	23.6	76.4	0.0	0.0	100.0
Pirojpur	Area	18.4	1.9	0	0	20.3
	%	90.6	9.4	0.0	0.0	100.0
Chittagong	Area	18.4	15.1	7	5.2	45.7
	%	40.3	33.0	15.3	11.4	100.0
Cox's Bazar	Area	7.2	16.2	17.3	14	54.7
	%	13.2	29.6	31.6	25.6	100.0
Noakhali	Area	6.3	39.9	3.4	0	49.6
	%	12.7	80.4	6.9	0.0	100.0
Laxmipur	Area	10.9	6.8	1.6	0	19.3
	%	56.5	35.2	8.3	0.0	100.0
Feni	Area	1.6	6.7	0.7	0	9
	%	17.8	74.4	7.8	0.0	100.0
Chandpur	Area	1.5	0	0	0	1.5
	%	100.0	0.0	0.0	0.0	100.0
Total	Area	835.41	1019.33	186.35	92.5	2133.45
Can't tell	4.29	7.01	2.50	1.21	1.25	3.57
Overall	100.00	100.00	100.00	100.00	100.00	100.00

Source: Karim, Z, S.G. Hossain and M. Ahmed (1990).

Table 2. Potential cropping patterns for different land types and varied soil salinity levels in Bangladesh

Salinity Category	Highland			Medium high land			Me dium lowland		
	Summer	Autumn	Winter	Summer	Autumn	Winter	Summer	Autumn	Winter
S_1 (2-4 ds m^{-1})	B.Aus rice Chilli Groundnut	L.T.Aman rice HYV Aman rice Orchard	Pulses Onion Watermelon Wheat Chilli Mustard Vegetables Sunflower	B.Aus ricc	L.T.Aman rice HYV Aman rice	Pulses Mustard Chilli Vegetable Wheat* Potato Cowpea Watermelon Sunflower	B.Aus rice	B.Aman rice L.T.Aman rice	HYV Boro rice Mustard Pulses Wheat Sweet potato
S_2 (4-8 ds m^{-1})	B.Aus rice Chilli Cucumber	L.T.Aman rice	Soybean Groundnut Cowpea Spinach Chilli Cucumber Mustard**	B.Aus rice	L.T.Aman rice HYV Aman rice	Mustard** Sesbenia Cowpea Soybean Chilli Groundnut Wheat* HYV Boro Sweet potato	B.Aus rice	B.Aman rice L.T.Aman rice	Soybean Mustard** Wheat* Spinach Sweet potato
S_3 (8-16 ds m^{-1})	Bermuda Grass Cucumber Cotton Sorghum	L.T.Aman rice	Wheat* Barley Cucumber Cotton Sorghum	Sesbania Sorghum	L.T.Aman rice	Cabbage Barley Cotton Wheat* Sorghum	Sesbania	L.T.Aman rice	Barley Cotton Wheat*
S_3 (>16 ds m^{-1})	Cotton Bermuda Grass *Leploc hioa fusca*		Cotton Barley	Cotton		Cotton	Cotton		Cotton Barley

*Where thermal situation permits and early sowing is possible
**Mustard variety, Daulut.
Source: Karim, Z, S.G. Hossain and M. Ahmed (1990).

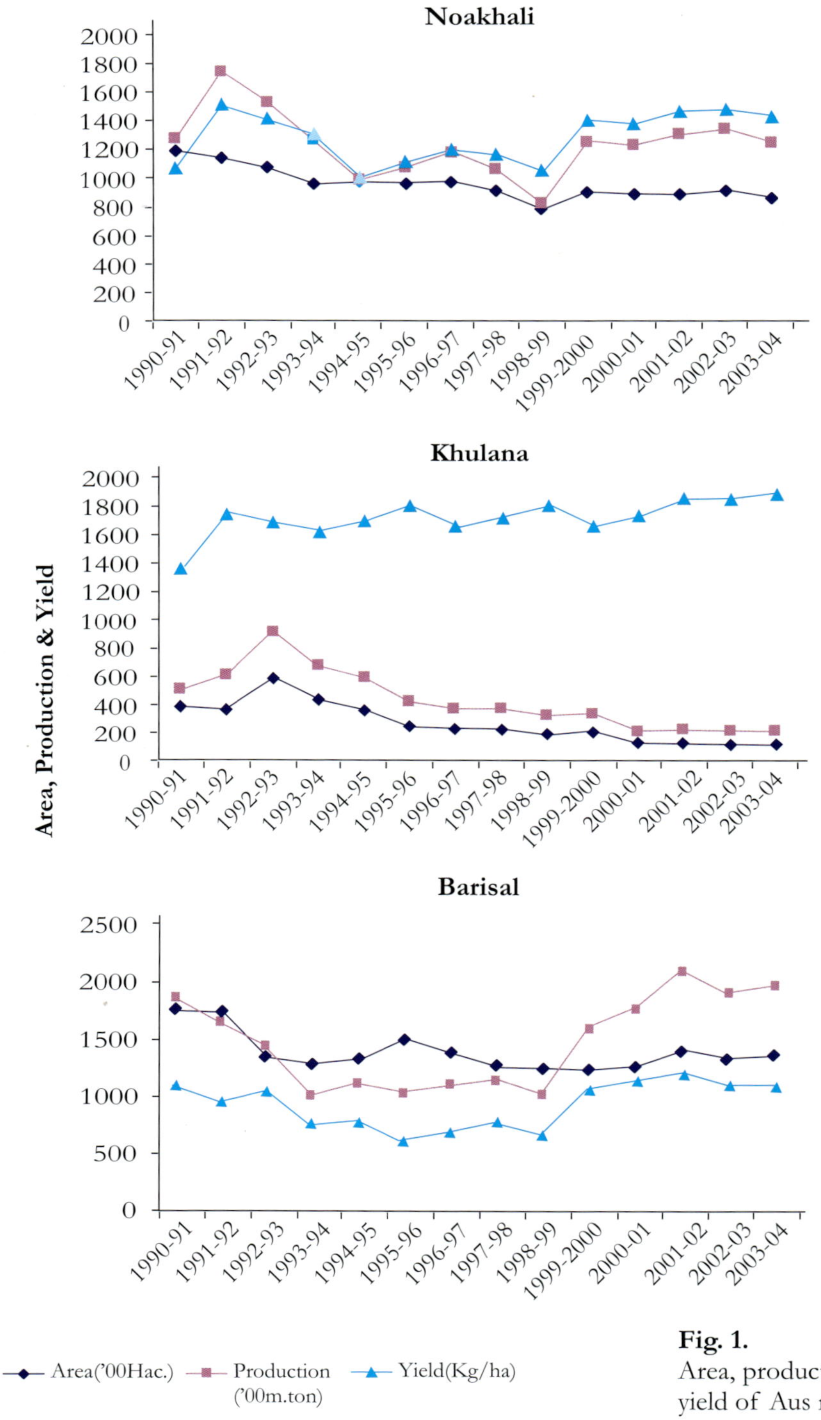

Fig. 1.
Area, production and yield of Aus rice

Variety cultivated and sources of seed

Nearly 100 percent of the salinity rice area of the sample farmers in Noakhali was covered by an HYV variety named BR-11. BR-11 is a popular rice variety in Noakhali because of its higher yield. On the other hand, 100 percent of the salinity rice area in Sathkhira was covered by an HYV variety named BR-41.

It was found that on average, 17 percent of rice seeds came from farmers' own seeds and 83 percent came from the market in production year 2005 in Noakhali . However, in the case of Sathkhira farmers, 50 percent of rice seeds came from farmers' own seeds and 50 percent came from the market (Figure 2).

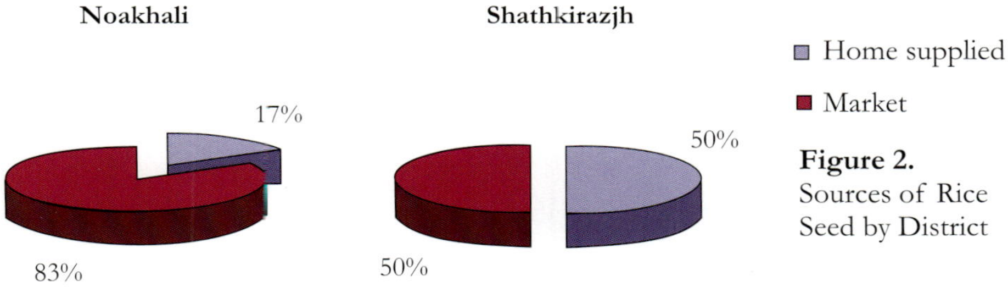

Figure 2. Sources of Rice Seed by District

Irrigation used in salinity affected rice field

In the salinity affected area, Aus rice is grown in the dry season and it requires irrigation. It was found that cost of irrigation (including labour) in Noakhali was $ 14.94/ha, and it was $ 54.66/ha in Sathkhira. Such application of irrigation resulted in an expenditure of 12.4 million US dollars in four districts of Khulna, Noakhali, Barishal and Patuakhali (Figure 6), indicating scope for cost savings in terms of reduction of irrigation for rice cultivation and for protecting the environment by means of adopting a salinity and drought resistant transgenic rice crop.

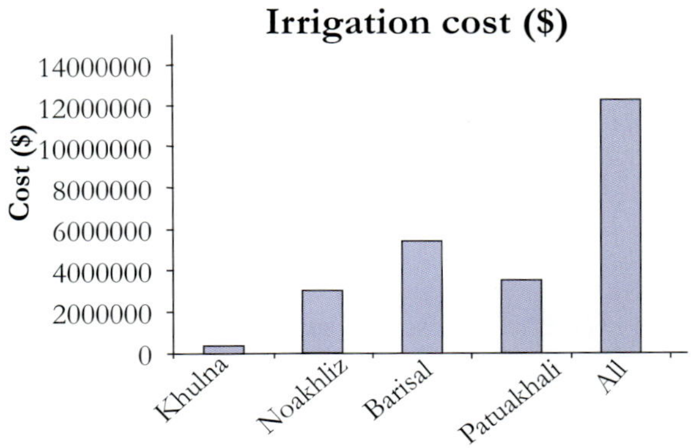

Figure 3.
Irrigation Expenditures in Aus Rice by Coastal Districts

Cost structure of rice production

The largest cost components of rice production in the salinity area of Noakhali were labour (54% of total variable costs) followed by fertilizer (18%) and draft power (12%) (Figure 4). The largest cost components of rice production in the salinity area of Sathkhira were labour (52% of total variable costs) followed by irrigation (16%) and draft power (13%) . In Noakhali, the share of irrigation cost covered 5 percent of total variable cost of rice production.

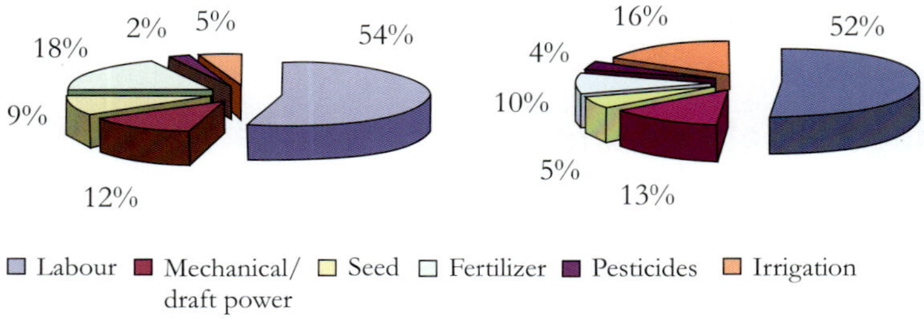

Figure 4.
Cost Structure of Aus Rice Cultivation in Soil Salinity Areas

Results of economic surplus analysis

Spreadsheets were used for estimating consumer surplus, producer surplus, total net benefits and rate of return. Research expenditures were assumed to begin in 2005 and to continue until 2009. Thus, there is a five years research, technology development, and regulatory lag before benefits occur. As noted above, the main assumptions for the economic surplus and benefit cost analyses were based on field surveys of farmers, scientists, and industry experts. The summary of assumed changes in the rice budget due to SDR rice are presented in Table 3

Yield of transgenic SDR rice is expected to increase by 20-40 percent. Irrigation cost is assumed to decrease by 25 percent, with slight increase in seed (10%), fertilizer (10%), and harvest labor costs. The probability of success is assumed to be 70-90 percent.

TABLE 3.
Summary of changes in transgenic salinity and drought resistant (SDR) rice budget (Based on farmer, scientists and industry expert opinions)

Particular	Changes
Variable cost	
Seed cost	Increase by 4-10%
Fertilizer cost	Increase by 5-10%
Irrigation cost	Decrease by 25%
Pesticide cost	No change
Labor cost	Increase by 2-6%
Fertilizer labor cost	
Harvest labor cost	Increase by 3% Increase by 4%
Other labor cost	no change
Other costs	no change
Yield & returns	
Yield (kg)	to increase by 20-40%
Price per kg	
Yield loss	10-75%
Variety	BR-4, 6, 28, 29, 30, 40, 41 and Tangail and Kajal Shahi
Probability of success	70-90%

Source: BADC and local market

TABLE 4.
Assumptions used in economic surplus model

Parameter	Description and Value
Year	Annual benefits are projected for 15 years after research commences, 2005 - 2019 (t = 1,2,……,15)
Supply elasticity	The supply elasticity, ε, is set at 0.13 for rice.
Demand elasticity	The demand elasticity, η, is set at- 0.20 for rice.
Proportionate yield change	The estimate of yield increase, E(*Y*), is 20%.
Proportionate change in input cost per hectare	The proportionate change in input cost per hectare is 14%.
Probability of research success	The technology has not been released yet, and the relevant probability of research success is set at 0.90.
Adoption rate	For the purpose of simulation, the assigned maximum adoption rates are 30%, 60%, and 70%. Time required to reach the peak: 3 years
Price	Wholesale prices for the period of 2000 - 2005 are averaged, giving a mean value of $ 183.98 per ton.
Quantity	The pre- research quantity is constant, equal to the base quantity of 323 thousand tons in 2005.
Research cost	The estimated annual research cost for SDR rice is $9538 thousand dollars.

Ex-ante analysis of cost and profitability of GM and non-GM SDR rice

The farmer survey revealed that salinity continues to be a major problem of rice production in Noakhali and Sathkhira causing annual losses in yield of 10 to 75 percent. Present crop regimes also require irrigation for rice production throughout the growing season. If an SDR rice were cultivated in Noakhali and Sathkhira, it would reduce irrigation costs per ha by $3.37, and $12.19 per ha (25 percent), respectively. On average, the per ha irrigation cost reduction would be $7.8. The commercialization of SDR rice would reduce labor cost in the salinity regions by 4 percent. The total variable cost saving in Noakhali for SDR rice would be 0.37 percent and in Sathkhira it would be 4.75 percent respectively (Table 5 and 6).

The yield of SDR rice is expected to increase by 460 kg/ha in Noakhali and by 603 kg/ha in Sathkhira. The minimum expected yield increase of GM SDR rice over Non-GM SDR rice is 20 percent. With this yield increase, the gross margin of SDR rice in Noahkhali would increase by 102 percent ($68/ha) and by 45 percent ($133/ha) in Sathkhira. However, in all these regions the extent of increase of gross margin of SDR rice over conventional non-GMO rice is about 49 percent (Table 7). The commercialization of a specific transgenic SDR rice variety could potentially offer a significant increase in gross margin and profitability to the producers

TABLE 5.
Ex-ante analysis of economic performance of SDR rice over Non- SDR rice in Noakhali

	Cost ($/ha)			
Cost Item	**Non- SDR rice**	**SDR rice**	**$ Change**	**% Change**
Labor	149.90	146.78	-4.62	-4.1
Mechanical/draft power	33.28	33.28	0.00	0
Seed	25.95	28.54	2.59	10.0
Fertilizer	49.04	53.94	4.90	10.0
Pesticides	6.36	6.36	0.00	0.0
Irrigation	13.47	10.10	-3.37	-25.0
Total variable cost	277.99	279.00	1.02	0.37
Yield(kg/ha)	2299.14	2759	459.8	0.2
Total return ($/ha)	344.16	413.00	68.83	20
Gross margin ($/ha)	66.18	133.99	67.82	102

TABLE 6.
Ex-ante analysis of economic performance of SDR rice over Non-SDR rice in Sathkhira

	Cost ($/ha)			
Cost Item	**Non- SDR rice**	**SDR rice**	**$ Change**	**% Change**
Labor	152.50	149.41	4.62	-4.0
Mechanical/draft power	38.77	38.77	0.00	0
Seed	15.65	17.22	1.56	10.0
Fertilizer	29.48	32.42	2.95	10.0
Pesticides	10.97	7.68	-3.29	-30.0
Irrigation	48.78	36.58	-12.19	-25.0
Total variable cost	277.99	279.00	1.02	0.37
Yield(kg/ha)	3016.13	3619	603.226	0.2
Total return ($/ha)	593.95	712.73	118.79	20
Gross margin ($/ha)	297.80	430.65	132.85	45

TABLE 7.
Ex-ante analysis of economic performance of SDR rice over Non-SDR rice in two regions

	Cost ($/ha)			
Cost Item	**Non- SDR rice**	**SDR rice**	**$ Change**	**% Change**
Labor	151.20	148.10	-4.62	-4.1
Mechanical/draft power	36.02	36.02	0.00	0
Seed	20.80	22.88	2.08	10.0
Fertilizer	39.26	43.18	3.93	10.0
Pesticides	8.45	8.45	0.00	0.0
Irrigation	31.12	23.32	-7.80	-25
Total variable cost	286.86	281.95	-4.91	-1.71
Yield(Kg/ha)	26110	31332	5222	0.2
Total return ($/ha)	2201.08	3133.20	932.12	42.35
Gross margin ($ha)	1914.22	2851.25	937.03	49

A of partial budget analysis of replacing traditional varieties by a SDR transgenic variety is presented in Table 7. The introduction of SDR rice would reduce irrigation cost by $7.80/ha and irrigation labor cost by $3.09/ha. The total incremental benefit is expected to be $947.89/ha against an incremental cost of $13.86/ha. Thus yielding a net benefit of $934.03/ha.

TABLE 8:
Partial budget of SDR rice replacing traditional varieties

Particular		Based on 2005 crop budget ($/ha)
Incremental benefits		
Reduced cost	Irrigation cost	7.80
	Irrigation labor	3.09
Added return		
	Increased revenue	937
Total incremental benefits		947.89
Incremental costs		
Added cost	Seed cost	2.08
	Fertilizer	3.93
	Harvest labor	7.85
Reduced return		–
Total incremental cost		13.86
Net Incremental benefits		**934.03**

Distribution of benefits and social returns

The results of economic surplus analysis model are presented in Table 9. The introduction of transgenic SDR rice in Bangladesh would have a high social return on investment. The total amount of discounted economic surplus from producing SDR transgenic rice over 10 years is projected at $302.8 million US dollars with a producer surplus of $184.1 million and a consumer surplus of $119.7 million if no international trade is assumed. If a small open economy is assumed, the projected benefits are similar in total, but would accrue to producers (some of them in the form of consumption). The internal rate of return (IRR) is estimated at 33.8 percent and a net present value (NPV) of benefits minus research costs is $215.7 million.

TABLE 9.
Ex-ante analysis of consumer and producer surplus, net benefits, and social rate of return from producing transgenic SDR rice in Bangladesh, 2010-2020

YEAR	Change in Economic surplus (thousand $)	Change in Consumer Surplus (thousand $)	Change in Producer surplus (thousand $)	Net BENEFIT (thousand $)	NPV (thousand $)	IRR (%)
2005	0.00	0.00	0.00	0.00	215,690.6	33.78
2006	0.00	0.00	0.00	0.00		
2007	0.00	0.00	0.00	0.00		
2008	0.00	0.00	0.00	0.00		
2009	0.00	0.00	0.00	0.00		
2010	0.00	0.00	0.00	0.00		
2011	23158.12	9122.90	14035.22	23158.12		
2012	47006.23	18517.61	28488.63	47006.23		
2013	55108.93	21709.58	33399.35	55108.93		
2014	55108.93	21709.58	33399.35	55108.93		
2015	55108.93	21709.58	33399.35	55108.93		
2016	55108.93	21709.58	33399.35	55108.93		
2017	55108.93	21709.58	33399.35	55108.93		
2018	55108.93	21709.58	33399.35	55108.93		
2019	55108.93	21709.58	33399.35	55108.93		
2020	55108.93	21709.58	33399.35	55108.93		
Total	511035.83	201317.15	309718.69	511035.83		
Discounted Total(at 5%)	303,818	119,686	184,132			

Sensitivity analysis

Sensitivity analysis was conducted under alternative scenarios of (a) a 30 percent increase in yield, (b) a 20 percent increase in input cost, (c) a 50 percent increase in supply elasticity, and (d) a 50 percent increase in base yield, cost, and supply elasticity. Under the various scenarios, the IRR ranges from 26 to 42 percent while the NPV ranged from $119 million to $367 million.

Environmental Impact

From our field surveys of farmers, scientists and industry experts, and also from farmers' focus group discussions, it is clear that the growers are aware about soil salinity problems of rice cultivation and most of them are risk averse, often trying to avoid risk of crop loss by irrigating. On the basis of our survey results, it was estimated that irrigation resulted in an expenditure of 12.4 million US dollars in salinity rice areas of four districts of Khulna, Noakhali, Barishal and Patuakhali (Figure 8), indicating there is a scope for cost savings in terms of reduction of irrigation for rice cultivation and for protecting the environment with conservation of ground water as well as solving soil salinity problems by means of adopting a salinity and drought-resistant transgenic rice crop.

Conclusion

Soil salinity is a major threat to irrigated agriculture. In Bangladesh, more than 30% of the net cultivated land is in the coastal area. Of the 2.85 million ha of coastal and off-shore lands, about 0.83 million ha of arable land are affected by varying degrees of soil salinity. Aus rice is the most important crop affected by salinity during the summer season in the coastal area of Bangladesh. Yield of salinity-affected Aus rice is very low and fluctuates from 1000 to 1500 kg/ha. In Khulna district, yield fluctuates from 1200 to 1600 kg/ha. In Barisal, yield is 1400 kg/ha. The minimum expected yield increase of GM SDR rice over Non-GM SDR rice is 20 percent. With this yield increase, the gross margin of SDR rice in Noahkhali would increase by 102 percent ($68/ha) and by 45 percent ($133/ha) in Sathkhira. However, in all these regions the extent of increase of gross margin of SDR rice over conventional non-GMO rice is about 49 percent. The commercialization of a specific transgenic SDR rice variety could potentially offer a significant increase in gross margin and profitability to the producers.

The introduction of SDR rice would reduce irrigation cost by $7.80/ha and irrigation labor cost by $3.09/ha. The total incremental benefit is expected to be $947.89/ha against an incremental cost of $13.86/ha, thus yielding a net benefit of $934.03/ha.

The introduction of transgenic SDR rice in Bangladesh would have sizable social benefits. The rate of social return or internal rate of return (IRR) is estimated at approximately 34% and a net present value (NPV) of $216 million. Environmental benefits should also be large.

Reference

Alston, J.M. , G.W. Norton, and P.G. Pardey (1995). Science under scarcity: Principles and practice for agricultural research evaluation and priority setting, Cornell University Press, Ithaca.

Alston, J.M, Hyde, J., Marra, M.C., & Mitchell, P.D. (2002). An ex ante analysis of the benefits from the adoption of corn rootworm resistant transgenic corn technology. *AgBioForum,* 5(3), 71-84. Available on the World Wide Web: http://www.agbioforum.org.

BBS (1996). Statistical Year Book of Bangladesh, Bangladesh Bureau of Statistics, Ministry of Planning, Government of People's Republic of Bangladesh, Dhaka.

Bhuian, S.I. (1997). Coping with water scarcity: what can rice science offer? Paper presented in the symposium on partnership in rice research for sustainable agricultural development in Bangladesh, BRRI, Gazipur, Bangladesh, 25 February 1997.

Commission of the European Communities. (2000). *Economic impacts of genetically modified crops on the agri-food sector.* Brussels: CEC Directorate-General for Agriculture. Available on the World Wide Web:http://europa.eu.int/comm/agriculture/publi/gmo/fullrep/ackn.htm.

Demont, M., and Tollens, E. (2004). First impact of biotechnology in the EU: Bt maize adoption in Spain. *Annals of Applied Biology,* 145, 197-207.

Demont, M., Wessler, J. and,Tollens, E. (2004). Biodiversity versus transgenic sugar beet: the one euro question. *European Review of Agricultural Economics,* 31, 1-18.

Dowley, L.J., Leonard, R., Rice, B., & Ward, S. (2001). *Low input fungicide programmes for the control of late blight in potatoes.* Oak Park, Carlow, Ireland: Teagasc Crops Research Centre. Available on the World Wide Web: http://www.teagasc.ie/research/reports/crops/4312/eopr-4312.htm.

Falck-Zepeda, J.B., Traxler, G., and R.G. Nelson, 2000. Surplus distribution from the introduction of a biotechnology innovation. American Journal of Agricultural Economics 82, 360-369.

Fulton, M., and Keyowski, L. (1999). The producer benefits of herbicide-resistant canola. *AgBioForum,* 2(2), 85-93. Available on the World Wide Web: http://www.agbioforum.org.

Ghosh, S.K. Majumder, RC, Chaudhuri, S.A. (1987). Studies on some oil seeds in Sunderban, Jr. of Indian Soc., Coast Agric. Res. 5(1). 165-167.

Karim, Z, S.G. Hossain and M. Ahmed (1990). Salinity problems and crop intensification in the coastal regions of Bangladesh, Soil publication No. 33, Soils and Irrigation Division, BARC, Farmgate, Dhaka 1215, Bangladesh, pp: 1-20.

Kalaitzandonakes, N.G. (2003). The economic and environmental impacts of agbiotech: An introduction. In N.G. Kalaitzandonakes (Ed.), *The Economic and environmental impacts of agbiotech* (pp.1-18). New York: Kluwer Academic.

May, M.J. (2003). Economic consequences for UK farmers of growing GM herbicide tolerant sugar beet. *Annals of Applied Biology,* 142, 41-48.

PG Economics. (2003). *Consultancy support for the analysis of the impact of GM crops on UK farm profitability.* Dorchester, Dorset, UK: PG Economics Ltd. Available on the World Wide Web: http://www.pgeconomics.co.uk/pdf/Cabinet_Office_GM_Crops.pdf

Cabinet_Office_GM_Crops.pdf. Master Plan Organization (MPO 1986). National water plan. MPO report, Vol. 1, Ministry of irrigation, Water Development and Flood Control, Dhaka 1000, Bangladesh, pp; 1-25.

Oster, J.D. (1994). Irrigation with poor quality water, J. Agric. Water Management, 25, 251-297.

Tolstrup, K., Andersen, S.B., Boelt, B., Buus, M., Gylling, M., Holm, P.B., et al. (2003). *Report from the Danish Working Group on the co-existence of genetically modified crops with conventional and organic crops.* Tjele, Denmark: Danish Institute of Agricultural Sciences. Available on the World Wide Web: http://web.agrsci.dk/gmcc-03/Co_exist_rapport.pdf

United States General Accounting Office. (2000). Biotechnology: *Information on prices of genetically modified seeds in the United States and Argentina* (GAO/RCED/NSIAD-00-55). Washington, DC: US General Accounting Office. Available on the World Wide Web: http://www.gao.gov/new.items/r400055.pdf.

Van Meijl, H., & Van Tongeren, F. (2003, August). *International diffusion of gains from biotechnology and the European Union's Common Agricultural Policy.* Paper presented at the 25th International Conference of Agricultural Economists, Durban, South Africa. Available on the World Wide Webhttp://www.iaaeagecon.org/conf/durban_papers/papers/064.pdf

Chapter **7**

Bt Eggplant for Fruit and Shoot Borer Resistant in Bangladesh

S.M. Fakhrul Islam and G.W. Norton

Introduction

Eggplant (*Solanum melongena*) is the second most important vegetable after potato in Bangladesh. It is extensively cultivated around homesteads and in commercial fields in both winter and rainy seasons. Eggplant covers an area of 148,330 ha, about 15% of the total vegetable area of Bangladesh, and its production was about 358,370 thousand tonness in 2003-04. Production of eggplant almost doubled from 1993-94 to 2003-04, while area increased to 148,330 ha in 2003-04 from 72,395 hectares in 1993-94 (Table 1).

Eggplant shoot and fruit borer (ESFB), *Leucinodes orbonalis* Guence is the most serious pest of eggplant (Alam and Sana, 1962, Chattopadhyay, 1987) in Bangladesh as well as in the Indian sub-continent (Dhankar, 1988). The yield loss caused by this pest has been estimated up to 67 percent in Bangladesh (Islam and Karim, 1991). The infestation ranged from 12-18 percent for shoot (Alam et al., 1964) and 20-86 percent for fruit in Bangladesh (Ali, et al., 1994). The shoot and fruit damage caused by this pest starts soon after transplantation and continues until the last harvest. The most common practice followed by the farmers for controlling this pest is spraying insecticide at an interval of 7-15 days or even more frequently which causes health and environmental problems.

Table 1.
Area, production and yield of eggplant in Bangladesh 1993-94 to 2003-04

Year	Area (ha)	Production (000 tonnes)	Yield (tonnes/ha)
1993-94	72,395	191,465	6.53
1994-95	72,475	188,135	6.41
1995-96	73,445	190,125	6.39
1996-97	74,335	192,620	6.40
1997-98	74,710	191,525	6.33
1999-00	160,690	392,340	6.03
2000-01	158,660	281,420	4.38
2001-02	155,640	378,390	6.01
2002-03	154,755	370,125	5.91
2003-04	148,330	358,370	5.97

Sources: BBS 1985-86, 1992, 1995 2000 and 2004

Economic information on genetically modified (GM) crops is available for major GM crop-growing countries (Commission of the European Communities, 2000; Dowley, et al., 2001; Falck-Zepeda, J.B., Traxler, G., and R.G. Nelson, 2000; Alston et al., 2002; Tolstrup, et al., 2003; Van Meijl and Tonegeren, 2003; Demont, et al., 2004a and 2004b: Alston, J. M, Hyde, J., Marra, M.C., & Mitchell, P.D., 2002). For Bangladesh, such information is not yet available. The aim of the present study is to assess the potential net economic benefits of ESFB resistant Bt eggplant in Bangladesh. The objective is to predict costs and benefits a producer would experience if a ESFB resistant Bt eggplant crop were to be cultivated in Bangladesh, as well as the aggregate benefits to society.

METHODS

Data

Given the intensity of eggplant cultivation, Narsingdi and Jamalpur districts were selected for field surveys for the present study. Data were collected from 60 farmers, 30 from each region. Data were also collected from 5 scientists and 8 industry experts from the Department of Agricultural Extension (DAE), the Bangladesh Agricultural Development Corporation (BADC), and seed companies.

Data collected from farmers included various input costs such as seed, fertilizers, pesticides and labor. Data were collected on crop varieties, sources of seeds, crop management practices, input and output prices, crop losses due shoot and fruit borer, and crop yields.

From scientists, data were collected on potential regions for Bt eggplant, expected yield changes if Bt eggplant were grown, changes in variable costs for Bt eggplant, time lags and costs for technology development and regulatory requirements.

From industry experts, data were collected on preferred varieties, sources of seed, extent of crop losses due to EFSB, expected extent of adoption of Bt eggplant, time path of adoption, and time lags and costs for technology development and meeting regulatory requirements. Economic surplus and benefit cost analysis

Cost and return information, secondary data on prices and quantities, and adoption information were combined in an economic surplus model to project total economic benefits and their distribution by region within the country and to consumers, producers, and seed sector (Alston, Norton, and Pardey 1995). A closed and a small open economy were used as alternative market assumptions. The costs of the research and product development (including meeting regulatory hurdles) were obtained from scientists and included along with the benefits in a benefit cost analysis of the public investment. Excel spreadsheets were used for estimating the economic surplus models and net present value of benefits over 15 years.

RESULTS AND DISCUSSION

Trend in eggplant acreage and yield

As indicated earlier in Table 1 and presented graphically below in Figure 1, eggplant acreage in Bangladesh increased from 72,395 ha in 1993-94 to 148,330 ha in 2003-04, with an annual growth rate of 10.5%. During this period, total eggplant production increased from 191,465 to 358,370 tonnes while yield declined slightly from 6.53 tonnes to 5.97 tonnes per ha. There was a positive trend in eggplant acreage in the selected two districts of Narsingdi and Jamalpur during 1993-94 to 2003-04 (Figure 2). Total eggplant acreage of Jamalpur increased from 10,470 ha to 10,690 ha with an annual average growth rate of 0.42 percent. The eggplant acreage of in Narsingdi also had a positive trend, increasing from 465 ha to 959 ha with an annual growth rate of 21 percent.

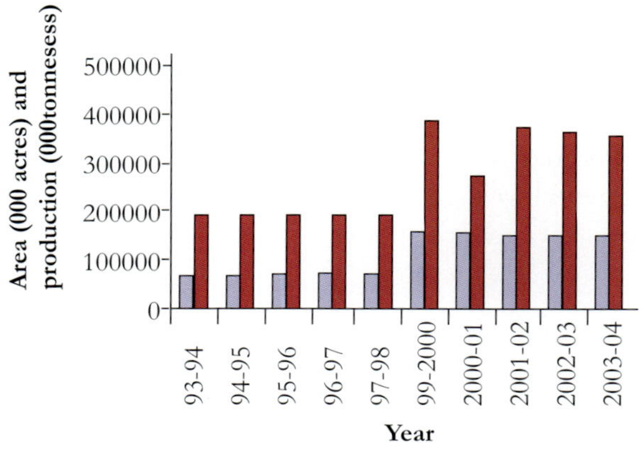

Area ■ Production

Figure 1.
Trend in area and production of eggplant in Bangladesh

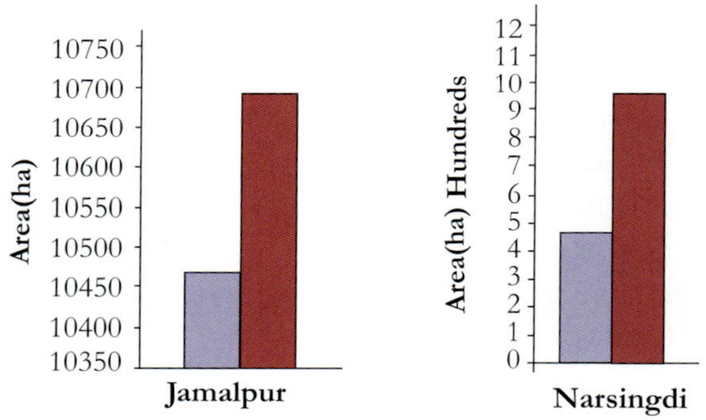

1998-99 Area

2003-04 Area

Figure 2.
Trend in area and production of eggplant in Bangladesh

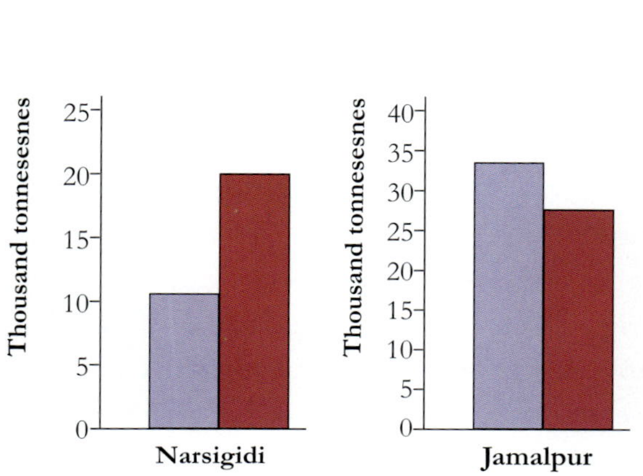

1998-99 Production ■ 2003-04 Production

Figure 3.
Trend in eggplant acreage in Jamalpur and Narsingdi

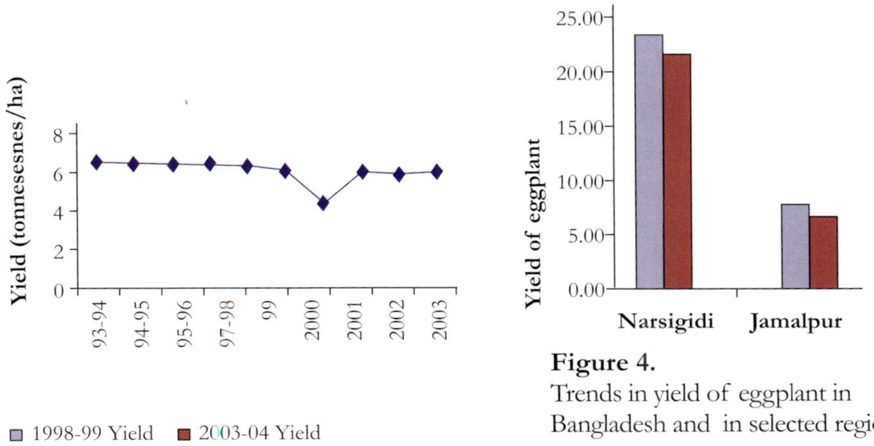

Figure 4.
Trends in yield of eggplant in Bangladesh and in selected regions

Eggplant production increased in all the districts as a result of area expansion during 1998-2004 (Figure 3). However, eggplant yield had a slightly decreasing trend in all the districts (Figure 4). On average, yield declined from 22.62 tonnes/ha to 20.67 tonnes/ha in Narsingdi. However, its yield was still higher than the national average yield of 6 tonnes/ha, while the yield of eggplant was almost stable in Bangladesh at about 6 tonnes/ha during 1993 to 2004, but much less than in Narsingdi. However, in Jamalpur, which is considered a leading intensive commercial eggplant growing area in Bangladesh, there was also a decreasing trend in eggplant yield, with 7.86 tonnes/ha in 1998-99 and 6.71 tonnes/ha in 2003-04.

Variety cultivated and sources of seed

It was found that on average 78% of eggplant seeds came from the farmers' own source and 22% from the market in 2005 in Narsingdi, and 100 percent of eggplant seed came from market in Jamalpur (Figure 5). There is a good potential market demand for shoot and fruit borer resistant eggplant seeds.

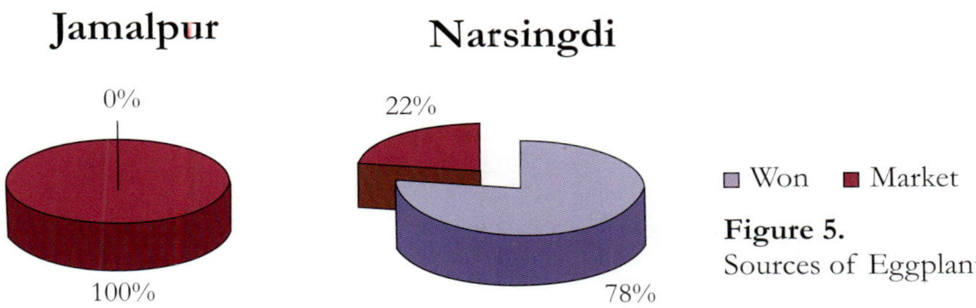

Figure 5.
Sources of Eggplant Seed

Insecticides used in eggplant production

Eggplant is attacked by as many as 53 species of insects (Nayar et al., 1995). Of these, eggplant shoot and fruit borer (ESFB), *Leucinodes orbonalis* Guence, is the most serious. The incidence of this pest occurs sporadically or in epidemic form every year throughout Bangladesh, affecting the quality and yield of crop. The damage caused by this pest in Bangladesh varies from 12-16% in shoot and 16-64% in fruits (Alam et al, 1964, Butani and Jotwani, 1994) and as a whole up to 70% loss is caused to the crop (Nair, 1986). The pest attack starts soon after transplanting and continues until the last harvest. For management of ESFB, Application of insecticides remains the primary method of control.

The study revealed that about 80% of total insecticides used in eggplant production was sprayed only for ESFB control in Narsindi and Jamalpur. On average, eggplant farmers of Narsingdi used 18.17 liter of liquid insecticides and 1.2 kg of solid insecticides per ha. In Jamalpur, use of liquid and solid insecticides in eggplant cultivation were 2.04 liter and 0.92 kg per ha, respectively (Figure 6). Considering these parameters, the total quantity of insecticides used in Narsingdi and Jamalpur for eggplant cultivation was estimated at 17,425 liters liquid plus 1150 kg solid and 21,808 liter liquid plus 9835 kg solid, respectively (Figure 7).

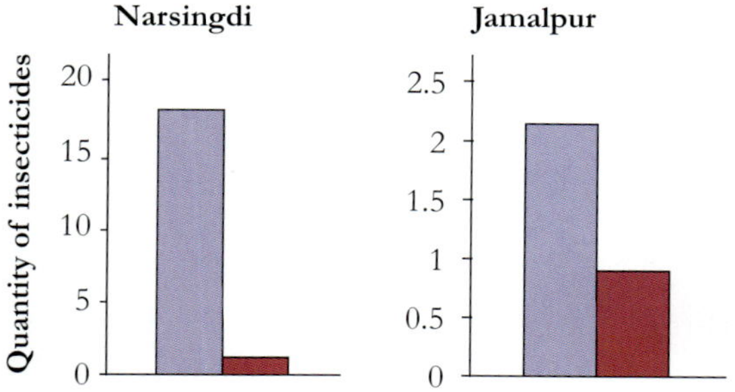

Fig 6.
Insecticides Used per ha in Eggplant Cultivation by District

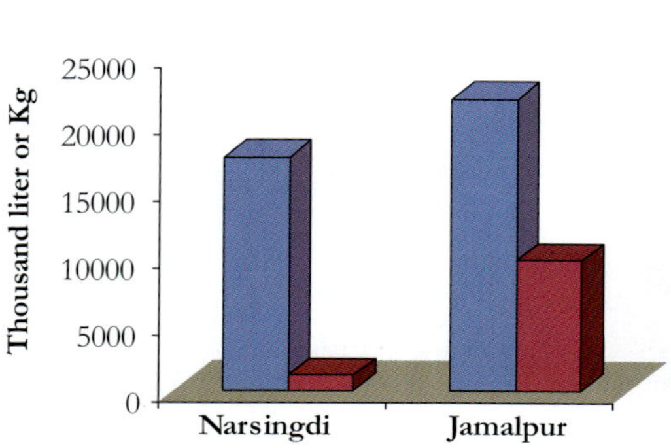

Figure 7.
Total insecticides Used in Eggplant Cultivation by District

Cost structure in eggplant production

The largest cost components in eggplant production in Narsingdi were human labor (61% of total variable cost), fertilizer (23%), and insecticides (5.2%) (Figure 8). These three inputs also had major cost shares in Jamalpur. In both areas, the share of insecticide cost was 4.6% of total variable costs of eggplant production.

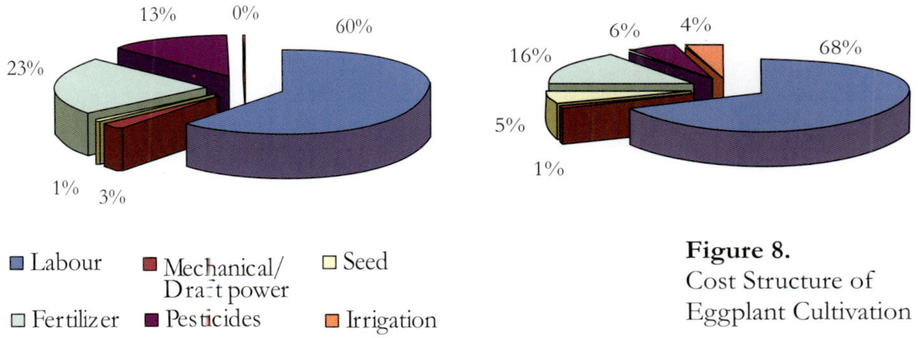

Figure 8.
Cost Structure of Eggplant Cultivation

Economic surplus analysis

Spreadsheets were used for estimating consumer surplus, producer surplus, net benefits and rates of return. Research and regulatory expenditures were assumed to begin in 2005 and to continue until 2009. The main assumptions for the economic surplus and rate of return analyses were based on field surveys of farmers, scientists and industry experts. A summary of assumed changes in the Bt eggplant budget is presented in Table 2. Yield of Bt eggplant is expected to increase by 15-30 percent. Pesticide cost is assumed to decrease by 70-90 percent, with slight increases in seed, fertilizer and harvest labor costs. The probability of success is assumed to be 70-90 percent.

TABLE 2.
Summary of changes in the eggplant budget due to adoption of Bt eggplant

Item	Change
Variable cost change	
Seeds cost	Increase by 4-10%
Fertilizer cost	Increase by 6-10%
Irrigation cost	No change
Pesticides cost	Decrease by 70-90%
Labor cost	Increase by 2-6%
Pesticide labor cost	Decrease by 70-75%
Harvest labour cost	Increase by 4-10%
Other labor cost	No change
Other costs	No change
Yield & returns	
Yield (kg)	Increase by 15-30%
Price per kg	No change
Gross return	Increase by 37-64%
Yield loss	15-30%
Variety	Singnath, Bottle
Probability of success	70-90%

Source: Private

TABLE 3.
Assumptions used in the economic surplus model for Bt eggplant

Parameter	Description and Value
Year	Annual benefits are projected for 15 years after research commences, 2005-2019 (t = 1,2,......,15)
Supply elasticity	The supply elasticity, ε ,is set at 0.5 for eggplant.
Demand elasticity	The demand elasticity, η, is set at 0.20 for eggplant.
Proportionate yield change	The estimated yield increase is 30%. Proportionate
change in input cost per hectare	The proportionate change in input cost per hectare is 0.36%
Probability of research success	The technology has not been released yet, and the estimated probability of research success is set at 0.80.
Adoption rate	For the purpose of simulation, the assigned maximum adoption rates are 30%, 40%, 50%, 60%, and 70%. Time required to reach the peak: 6 years
Price	Wholesale prices for the period of 2000-2005 are averaged, giving a mean value of $ 123.08 per tonneses.
Quantity	The pre-research quantity is constant, equal to the base quantity 536 thousand tonneses in 2005.
Research cost	The estimated annual research cost for eggplant is $15,384.62 per year.

Analysis of cost and profitability of GM and non-GM eggplant

Fruit and shoot borer continues to be a major problem in Narsingdi and Jamalpur as well as in other regions of Bangladesh, causing annual yield losses of 10 to 15 percent and high rates of pesticide applications. If a Bt eggplant were to be cultivated in Narsingdi and Jamalpur, it is projected to reduce insecticide cost per ha by $100.54, and $25.10, respectively. In all regions, per ha insecticide cost reduction would be $70.36 or 75% of the total cost of pesticides. The commercialization of a specific transgenic SFBR eggplant variety could potentially offer considerable cost savings to producers (Gianessi, Sankula, & Reigner, 2003). The survey revealed that it would increase labor cost in two regions by 4%, seed cost by 7%, and fertilizer cost by 6%. Total variable cost savings in Narsingdi for Bt eggplant would be $0.40/ha (0.02%) and in Jamalpur $21.46/ha (1.67%) (Table 4 and 5).

The yield of Bt eggplant would be expected to increase by 8277 kg/ha in Narsingdi and by 12,051 kg/ha in Jamalpur. The minimum expected yield increase of Bt eggplant over Non-Bt eggplant is 30%. With this yield increase, the gross margin of Bt eggplant in Narsingdi would increase by $2030.85/ha (46.5%) and by $1590.12 (41%) in Jamalpur. However, in all regions the extent of increase of gross margin of Bt eggplant over conventional non-BT eggplant would be about 44.8% (Table 6). The commercialization of a specific transgenic FSBR eggplant variety could potentially offer a significant increase in gross margin and profitability to the producers.

TABLE 4.
Ex-ante analysis of economic performance of Bt eggplant and non-Bt eggplant in Narsingdi region.

Cost Item	Non- Bt Eggplant	Bt Eggplant	Change	% Change
Labor	1586.00	1649.44	63.44	4.0
Mechanical/draft power	72.24	72.24	0.00	0
Seed	17.86	19.11	1.25	7.0
Fertilizer	590.79	626.23	35.45	6.0
Insecticides	134.05	33.51	-100.54	-75.0
Irrigation	10.08	10.08	0.00	0
Total variable cost	2411.02	2410.62	-0.40	-0.02
Yield (kg/ha)	27589	35866	8277	30.0
Total return ($/ha)	6775.93	8806.38	2030.45	30.0
Gross margin ($/ha)	4364.91	6395.76	2030.85	46.5

TABLE 5.
Ex-ante analysis of economic performance of Bt egg plant and non-Bt eggplant in Jamalpur region

Cost Item	Non- Bt EggplantBt	Eggplant	Change	% Change
Labour	1022.57	1063.47	40.90	4.0
Mechanical/draft power	20.17	20.17	0.00	0
Seed	80.76	86.41	5.65	7.0
Fertilizer	240.02	254.42	14.40	6.0
Insecticides	33.46	8.37	-25.10	-75.0
Irrigation	62.41	62.41	0.00	0
Total variable cost	1459.39	1480.85	21.46	1.47
Yield (kg/ha)	40171	52223	12051	30
Total return ($/ha)	5362.14	6973.72	1611.58	30.1
Gross margin ($/ha)	3902.75	5492.88	1590.12	40.7

TABLE 6.
Ex-ante analysis of economic performance of BT eggplant and non-Bt eggplant in all regions

Cost Item	Non- Bt EggplantBt	Eggplant	Change	% Change
Labour	1359.41	1413.79	54.38	4.0
Mechanical/draft power	51.42	51.42	0.00	0
Seed	43.02	46.03	3.01	7.0
Fertilizer	458.54	464.17	5.63	6.0
Insecticides	93.82	23.45	-70.36	-75.0
Irrigation	31.01	31.01	0.00	0
Total variable cost	2037.22	2029.87	-7.35	-0.36
Yield (kg/ha)	32621.83	42408	9786.549	30
Total return ($/ha)	6210.42	8070.64	1860.22	30.0
Gross margin ($/ha)	4173.20	6040.77	1867.57	44.8

A partial budget analysis of replacing a traditional variety with a transgenic one is presented in Table 7. The introduction of SFBR eggplant would reduce insecticide cost by $36.36/ha and insecticide labor cost by $34/ha. The total incremental benefit is estimated to be $1930.58/ha against an incremental cost of $62.39/ha, yielding a net benefit of $1868.19/ha.

TABLE 7:
Partial budget of Bt eggplant replacing traditional varieties

Particular		Based on 2005 crop budget ($/ha)
Incremental benefits		
Reduced cost		
	Insecticides cost	36.36
	Insecticides labor	34
Added return		
	Increased revenue	1860.22
Total incremental benefits	1930.58	
Incremental costs		
Added cost		
	Seed cost	3.01
	Fertilizer	5.63
	Harvest labor	53.75
Reduced return		-
Total incremental cost		62.39
Net Incremental benefits		1868.19
Incremental Benefit Cost ratio		31

Distribution of benefits and social return

The results of the economic surplus analysis are presented in Table 8, assuming a closed economy. The introduction of Bt eggplant in Bangladesh would have a high social return on investment. Table 8 shows the estimated social return and its distribution between producers and consumers for the period 2110 to 2119. The internal rate of return (IRR) is estimated at 17.8% and the net present value (NPV) at $65.5 million.

TABLE 8.
Projected amount of consumer and producer surplus, net benefits, and social rate of return from producing Bt eggplant in Bangladesh, 2010-2019

YEAR	Change in Economic surplus (thousand $)	Change in Consumer Surplus (thousand $)	Change in Producer surplus (thousand $)	Net BENEFIT (thousand $)	NPV (thousand $)	IRR (%)
2005	0.00	0.00	0.00	-11000.00	65,544.46	17.77
2006	0.00	0.00	0.00	-11000.00		
2007	0.00	0.00	0.00	-11000.00		
2008	0.00	0.00	0.00	-11000.00		
2009	0.00	0.00	0.00	-11000.00		
2010	9673.09	6045.68	3627.41	9673.09		
2011	12954.97	8096.86	4858.12	12954.97		
2012	16265.63	10166.02	6099.61	16265.63		
2013	19605.04	12253.15	7351.89	19605.04		
2014	19605.04	12253.15	7351.89	19605.04		
2015	22973.22	14358.26	8614.96	22973.22		
2016	22973.22	14358.26	8614.96	22973.22		
2017	22973.22	14358.26	8614.96	22973.22		
2018	22973.22	14358.26	8614.96	22973.22		
2019	22973.22	14358.26	8614.96	22973.22		

Sensitivity analysis

Sensitivity analysis was conducted under alternative scenarios of (i) a 20 percent increase in yield, (ii) a 20 percent increase in input cost, (iii) a 50 percent change in supply elasticity, and (iv) a 50 percent increase in base yield, cost, and supply elasticity. Under the first alternative scenario of a 20 percent increase in yield, compared to the base situation, the IRR and NPV decreased from 17.8% to 11.4% and from $65.5 million to $27.37 million dollar, respectively. In the case of 20% increase in input costs, the Bt technology would remain viable with a 9% IRR and an NPV of 35.6 million dollars. In the case of 50 percent increase in the supply elasticity, the IRR decreased from 17.8% to 11.5% and NPV from $65.5 million to $27.5 million. Consumer surplus would decrease by 24.1 percent and producer surplus would decrease by 49.5 percent. In the case of a 50 percent increase in base yield, input cost and supply elasticity, the IRR, NPV, net benefit, and producer and consumer surplus only changed slightly.

Environmental Impact

From our field surveys of farmers, scientists and industry experts, and also from farmers' focus group discussions, it is clear that the growers are aware about shoot and fruit borer and most of them are risk averse, often trying to avoid risk of crop loss by spraying high doses in advance of shoot borer attack. The frequency of spraying is often 2 to 3 times per week. On the basis of our survey results, it was estimated that total pesticide use in eggplant production in Bangladesh is 703,817 liters and 65,457 kg solids which can be greatly reduced (80%) by using Bt eggplant. Thus, the extent of environmental pollution is much higher in the case of non-Bt eggplant. Pesticide use is causing pollution of waterways and groundwater, harm to non-target organism and species biodiversity. Pesticide applications are causing farmers a health hazard, killing beneficial insects, and destroying our flora and fauna. Moreover, it has a residual effect in the food chain. Therefore, a safer Bt technology is desirable for protection of health, environment, and biodiversity. The argument is that because Bt eggplant results in the use of fewer, less toxic, and persistent insecticides, by implication, it should also lead to decrease in negative impacts on human health and non-target organisms.

Conclusion

Eggplant is the second most important vegetable after potato grown in Bangladesh. It is extensively cultivated in homestead and commercial fields in both winter and rainy seasons. Eggplant shoot and fruit borer is considered to be most important problem in its cultivation in Bangladesh. Farmers use a large amount of chemical pesticides to control it.

The study revealed that about 80 percent of total insecticides used in eggplant production were sprayed only for ESFB control in Narsingdi and Jamalpur. On average, the eggplant farmers of Narsingdi used 18.17 liter of liquid insecticides and 1.2 kg of solid insecticides per ha. In Jamalpur per ha use of liquid and solid insecticides in eggplant cultivation were 2.04 liter and 0.92 kg, respectively. Such a large quantity of insecticides indicates that there is a good scope of cost savings in terms of insecticide use for eggplant cultivation, and for protecting the environment by means of adopting ESFB transgenic eggplant.

The minimum expected yield increase of Bt eggplant over Non-Bt eggplant is 30%. With this yield potential, the gross margin of Bt eggplant in Narsingdi would increase by $2030.85/ha (30%) and by $1590.12/ha (22%) in Jamalpur. However, in all regions the extent of increase of gross margin of Bt eggplant over conventional non-BT eggplant is about 44.8%. The commercialization of a specific transgenic FSBR eggplant variety could potentially offer a significant increase in gross margin and profitability to the producers.

Partial budget analysis of replacing a traditional variety by ESFB transgenic showed that introduction of SFBR eggplant would reduce insecticide cost by $36/ha and insecticide labor cost by $34/ha. The total incremental benefit was found to be $1930/ha against an incremental cost of $62/ha, yielding a net benefit of $1868/ha.

Introduction of transgenic SFBR eggplant in Bangladesh would have sizable potential benefits. The internal rate of return is estimated at 17.8% and the net present value at $65.5 million. Results of sensitivity analysis implied that the transgenic SFBR eggplant technology would remain viable over the exiting non-transgenic one in the economy under a wide range of assumed parameters.

Reference

Alam, M, Z, and D.L. Sana, (1962). Biology of the brinjal shoot and fruit borer, *Leucinodes orbonalis* Guen. (Pyralidae: Lepidoplera) in East Pakistan. The Scientist S(1-4): 13-124.

Alam, M. Z., A. Ahmad, S. Alam and M. A. Islam (1964). A Review of Research Division of Entomology (1947-1964), Agricultural Information Service, Dhaka, pp. 77.

Ali, M. I (1994). Natural enemies recorded from brinjal fruit and shoot borer, *Leucinodes orbonalis* Guence durin 1992-94. pp. 119-120, In{Annual Research Repot, 1993-94. Entomology Division, BARI, Gazipur, Bangladesh.

Alston, J.M., G.W. Norton, and P.G. Pardey (1995). Science under scarcity: Principles and practice for agricultural research evaluation and priority setting, Cornell University Press, Ithaca.

Alston, J.M, Hyde, J., Marra, M.C., & Mitchell, P.D. (2002). An ex ante analysis of the benefits from the adoption of corn rootworm resistant transgenic corn technology. AgBioForum, 5(3), 71-84. Available on the World Wide Web: http://www.agbioforum.org.

BBS (1985, 86, 1992, 1995, 2000, and 2004). Statistical Year Book of Bangladesh (Various issues), Bangladesh Bureau of Statistics, Ministry of Planning, Government of People's Republic of Bangladesh, Dhaka.

Chattopadhyay, P. (1987). Entomology pest control and crop protection. West Bengal State Board, Arjo Mension (9th floor) 6A Raja Skubodh Mollick Square, Calcutta 700013, India. Pp- 304.

Commission of the European Communities. (2000). Economic impacts of genetically modified crops on the agri-food sector. Brussels: CEC Directorate-General for Agriculture. Available onthe World Wide Web http://europa.eu.int/comm/agriculture/publi/gmo/fullrep/ackn.htm

Demont, M., and Tollens, E. (2004a). First impact of biotechnology in the EU: Bt maize adoption in Spain. *Annals of Applied Biology*, 145, 197-207.

Demont, M., Wessler, J. and,Tollens, E. (2004b). Biodiversity versus transgenic sugar beet: the one euro question. *European Review of Agricultural Economics,* 31, 1-18.

Dhankar, B.S. (1988). Progress in resistance studies in eggplant *(Solanum melongena, L)* against shoot and fruit borer *(Leucinodes orbonalis Guen.)* infestation , Tropical Pest Management, 34: 343-345.

Dowley, L.J., Leonard, R., Rice, B., & Ward, S. (2001). *Low input fungicide programmes for the control of late blight in potatoes.* Oak Park, Carlow, Ireland: Teagasc Crops Research Centre. Available on the World Wide Web: http://www.teagasc.ie/research/reports/crops/4312/eopr-4312.htm

Falck-Zepeda, J.B., Traxler, G., and R.G. Nelson, 2000. Surplus distribution from the introduction of a biotechnology innovation. American Journal of Agricultural Economics 82, 360-369.

Islam, M. N. and M. A. (Karim, 1991). Management of brinjal shoot and fruit borer, *Leucinodes orbonalis* Guence (Lepidoptera: Pyralidae) in field. In annual Research Report, 1990-91, Entomology Division, BARI, Gazipur, Bangladesh, pp.-44-46.

Tolstrup, K., Andersen, S.B., Boelt, B., Buus, M., Gylling, M., Holm, P.B., et al. (2003). *Report from the Danish Working Group on the co-existence of genetically modifed crops with conventional and organic crops.* Tjele, Denmark: Danish Institute of Agricultural Sciences. Available on the World Wide Web: http://www.teagasc.ie/research/reports/crops/4312/eopr-4312.htm

Van Meijl, H., & Van Tonegeren, F. (2003, August). *International diffusion of gains from biotechnology and the European Union's Common Agricultural Policy.* Paper presented at the 25th International Conference of Agricultural Economists, Durban, South Africa. Available on the World Wide Web: http://www.teagasc.ie/research/reports/crops/4312/eopr-4312.htm

Chapter **8**

Late Blight Resistant Potato in Bangladesh

S.M. Fakhrul Islam and G.W. Norton

Introduction

Potato is the third most important crop grown in Bangladesh during the winter season (Baset, 2003). Production of potato has increased substantially during the past two decades. Area under potato production has increased to more than 245 thousand ha and production has tripled since 1980. Average yield per hectare has also increased from 10 to 12 tonnes per ha (Table 1).

The Tuber Crops Research Center (TCRC) of BARI has conducted potato research in Bangladesh since independence. Its research is focused mainly on varietal development, agronomic management and post-harvest issues. Since 2000, BARI has released 17 modern potato varieties. Two popular BARI potato varieties are Diamont and Cardinal. About 45 percent of the potato area is covered by Diamont, 20 percent by cardinal, and 33 percent by local varieties (Baset, 2003).

Late blight (*Phytophthora infestans*) is considered the most important problem of potato cultivation in Bangladesh. Farmers apply large amounts of chemical pesticides to control this disease, increasing the cost of cultivation and creating potential health and environmental problems. Each year, crop losses due to this disease are around is 20 to 30 percent.

The purpose of the present study is to provide an assessment of the potential economic impacts of transgenic late blight disease resistant (LBR) potato in Bangladesh. The objective is to predict costs and/or benefits that a producer might experience if a transgenic LBR potato were to be cultivated in Bangladesh, as well as the aggregate benefits to society.

TABLE. 1.
Area, yield and production of potato in Bangladesh 1981-82 to 2002-2003

Year	Area ('000 ha)	Yield (tonnes/ha)	Production ('000 tonnes)
1981-82	107.71	10.10	1087.80
1982-83	110.13	10.27	1131.09
1983-84	110.46	10.39	1147.69
1984-85	111.33	10.25	1141.02
1985-86	108.49	10.01	1085.33
1986-87	106.49	10.04	1069.31
1987-88	123.04	10.37	1275.65
1988-89	111.36	9.78	1089.36
1989-90	116.63	9.14	1065.68
1990-91	123.90	9.98	1236.81
1991-92	127.85	10.79	1379.37
1992-93	129.66	11.09	1438.02
1993-94	131.30	10.95	1438.06
1994-95	131.65	11.15	1468.47
1995-96	132.24	11.28	1491.56
1996-97	134.03	11.25	1507.87
1997-98	136.33	11.39	1553.18
1998-99	244.94	11.28	2762.00
1999-00	243.52	12.05	2933.32
2000-01	249.99	12.86	3215.57
2001-02	237.60	12.60	2994.00
2002-03	245.30	13.80	3386.00

Sources: Bangladesh Bureau of Statistics 1985-86, 1992, 1995 and 2000, 2004

METHODS

Data

Given the intensity of potato cultivation, Rangpur, Bogra and Munshiganj districts were selected for field surveys for the present study. Data were collected from 90 farmers, 30 from each region. Data were also collected from 5 scientists and 8 industry experts from the Department of Agricultural Extension (DAE), the Bangladesh Agricultural Development Corporation (BADC), and seed companies.

Data collected from farmers included input costs such as seed, fertilizers, pesticides and labor. Data were collected on crop varieties, sources of seeds, crop management practices, input and output prices, crop losses due to late blight, and crop yields.

From scientists, data were collected on potential regions for GM potatoes, expected yield changes if GM potato were grown, changes in variable costs for GM potato, time lags and costs for technology development and regulatory requirements.

From industry experts, data were collected on preferred varieties, sources of seed, extent of crop losses due to late blight, expected extent of adoption of GMO potato, and time path of adoption, time lags and costs for technology development and regulatory requirements.

Economic surplus and benefit cost analysis

Cost and return information, secondary data, and adoption information were combined in an economic surplus model to project total economic benefits and their distribution by region within the country and to consumers, producers, and seed sector. (Alston, Norton, and Pardey 1995). A closed economy model was applied. The costs of the research and product development (including meeting regulatory hurdles) were included along with the benefits in a benefit cost analysis of the public investment. Excel spreadsheets were used for estimating the economic surplus models.

RESULTS AND DISCUSSION

Trend in potato acreage and yield

As described in Table 1, there was a positive trend in potato acreage in the three selected districts of Rangpur, Bogra and Munshiganj during 2000-01 to 2004-05 (Figure 1). There was a positive trend in potato acreage in all of the eight upaupazillas of Rangpur District during this period. Total potato acreage of Rangpur district increased from 30,630 ha to 39,160 ha with an annual average growth rate of 6 percent. The potato acreage in Bogra also had a positive trend, increasing from 30,492 ha to 40,500 ha with an annual growth rate of 6.5 percent. The potato acreage of Munshiganj increased from 28,327 ha in 2000-01 to 32,232 ha in 2004-05, with an annual growth rate of 3.5 percent.

Potato acreage in various Upazilla of Rangpur (ha)

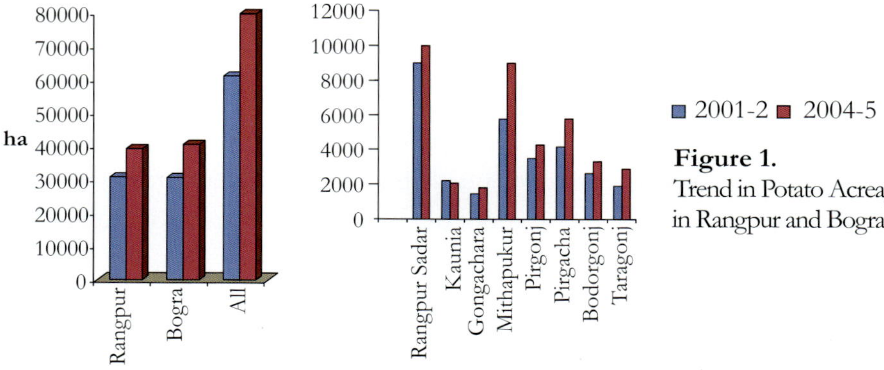

Figure 1.
Trend in Potato Acreage in Rangpur and Bogra

Potato production increased in all the districts as a result of expansion of area during 2001-2005 (Fig. 2). However, potato yield decreased in all the upazillas of Rangpur. On average, yield declined from 20.5 tonnes/ha to 17 tonnes/ha in Rangpur. However, its yield was still higher than the national average yield of 12 tonnes/ha while yield of potato was almost stable in Bogra at 13.7 tonnes/ha and 13.9 tonnes/ha during 2001-2 to 2004-5, respectively. The farmers of Rangpur believe that their potato yield has declined over the years as a result of degradation of an old variety. However, in Munshiganj, which is considered a leading commercial potato growing area in Bangladesh, there has been an increasing trend in potato yield. Its yield increased from 27.64 tonnes/ha in 2001-02 to 28.30 tonnes/ha in 2004-05.

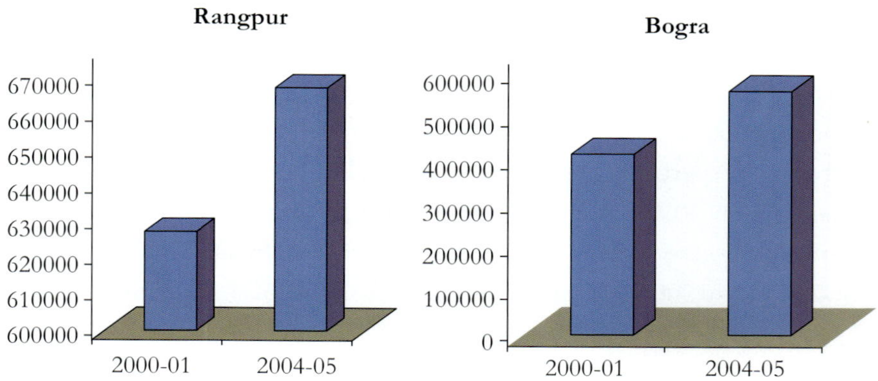

Figure 2.
Trend in Potato Production in Rangpur and Bogra

Variety cultivated and sources of seed

Eighty three percent of the potato area in Rangpur was covered by HYVs and 17 percent by local varieties (Fig. 3). The popular varieties in Rangpur were Diamont, Cardinal and Lal Pakhri. Recently a new variety, Granula, is also gaining popularity because of its higher yield. In Bogra, 58 percent of the potato area was covered by HYVs and 42 percent by local varieties. Diamont, Cardinal, and Lal Pakhri were also the popular potato varieties in Bogra. The most popular variety in Munshiganj was Diamont, which covered around 95 percent of the potato area, while Binola covered 3 percent and others 2 percent.

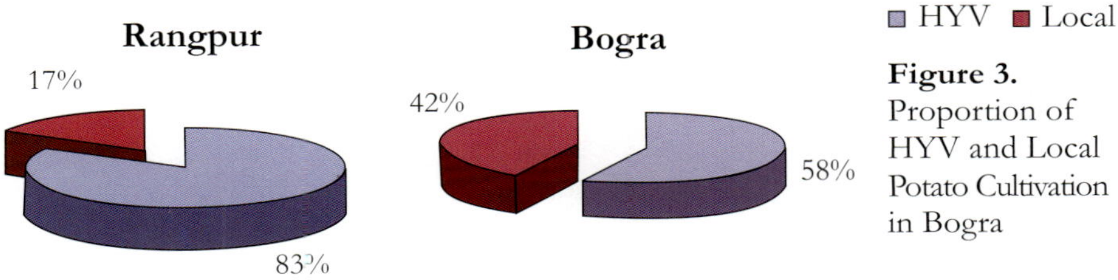

Figure 3. Proportion of HYV and Local Potato Cultivation in Bogra

On average, 53 percent of potato seeds came from farmers' own seeds and 45 percent came from the market in production year 1994 (Figure 4). However, in the case of medium and large farmers, the market was the largest source of potato seed (84 percent).

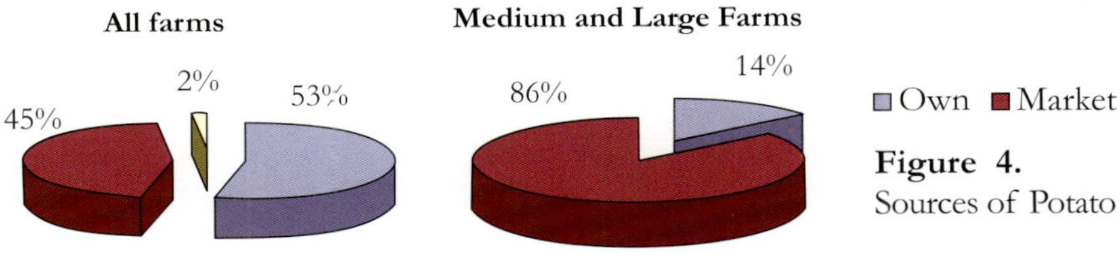

Figure 4. Sources of Potato Seed

Pesticides used in potato production

The study found that about 85 percent of total pesticides used in potato production were for late blight control in Rangpur and the figure was similar (80 percent) in Bogra. On average, potato farmers in Rangpur used 3.6 liters of liquid pesticides and 4.1 kgs of solid pesticides. In Bogra, per ha use of liquid and solid pesticides in potato cultivation were 5.7 liter and 4.9 kg, respectively, while farmers of Munshiganj used more solid pesticides (9 kg/ha) and less liquid (2 liter/ha). Considering these parameters, the total quantity of pesticides applied in Rangpur and Bogra for potato cultivation was estimated at 85 thousand liquid liters plus 106 thousand kgs solid and 58 thousand liquid liters plus 121.5 thousand kg solid, respectively. Such a huge quantity of pesticides resulted in an expenditure of $56 million dollars in three districts, indicating scope for cost savings in pesticides environmental improvement by means of adopting a late blight resistant transgenic potato crop.

Cost structure in potato production

The largest cost components of potato production in Rangpur were seeds (36% of total variable costs) labour (26%), and fertilizers (22%). These three inputs also covered most of the cost shares in Bogra. In both areas, the share of pesticides cost covered 5 percent of total variable cost of potato production.

Economic Surplus Analysis

Spreadsheets were used to estimate consumer surplus, producer surplus, total net benefits and rate of return. Research expenditures were assumed to begin in 2005 and to continue until 2009. There is a five years research, technology devlopment, and regulatory lag before benefits occur. The main assumptions for the economic surplus and rate of return analysis were based on field surveys of farmers, scientists, and industry experts as noted above. A summary of assumed changes in the potato budget due to LBR potato is presented in Table 2. Yield of transgenic potato is expected to increase by 15-30 percent. Pesticide cost is assumed to decrease by 70-90 percent, with increases in seed, fertilizer, and harvest labor costs. The probability of success is assumed to be 70-90 percent.

TABLE 2:
Summary of changes in Bt potato budget due to adoption of LBR potato

Item	Change
Variable cost change	Decrease by 17%
Seeds cost	Increase by 4-10%
Fertilizer cost	Increase by 6-10%
Irrigation cost	No change
Pesticides cost	Decrease by 70-90%
Labor cost	Increase by 2-6%
Pesticide labor cost	Decrease by 70-75%
Harvest labour cost	Increase by 4-10%
Other labor cost	No change
Other costs	No change
Yield & returns	
Yield (kg)	Increase by 15-30%
Price per kg	No change
Gross return	Increase by 37-64%
Current Yield loss	15-30%
Variety	Cardinal, Diamont
Probability of success	70-90%

Source: BADC, Private

Analysis of costs and profitability of GM and non-GM potato

According to our farmer survey, late blight continues to be a major problem in Rangpur, Bogra, and Munshiganj causing annual losses in yield (10 to 15 percent). Present crop regimes require a regular, high rate of fungicide application at short intervals throughout the growing season. If an LBR potato is cultivated in Rangpur, Bogra, and Munshiganj, it would reduce fungicide cost per ha by $30.7, $47.2 and $51.2 per ha (80 percent), respectively. On average, the per ha pesticide cost reduction would be $43.5 or 85 percent of total pesticide cost. The commercialization of LBR potato would reduce labor cost in the three regions by 2 percent. The total variable cost saving in Rangpur for LBR potato would be 0.24 percent and in Bogra and Munshiganj 1.4 and 0.9 percent respectively (Table 3, 4 and 5).

The yield of LBR potato is expected to increase by 3593 kg/ha in Rangpur, by 3876 kg/ha in Bogra, and by 4374 kg/ha in Munshiganj. The minimum expected yield increase of GM potato over Non-GM potato is 15 percent. With this yield increase the gross margin of LBR potato in Rangpur would increase by 38 percent ($279/ha), by 30 percent ($284/ha) in Bogra, and by 40 percent ($459.5/ha) in Mushiganj. However, in all regions the extent of increase of gross margin of LBR potato over conventional non-GMO potato is about 37 percent (Table 6). The commercialization of a specific transgenic LBR potato variety could potentially offer a significant increase in gross margin and profitability for producers.

TABLE 3.
Analysis of economic performance of GMO potato and non-GMO potato in the Rangpur region

Cost Item	Cost ($/ha)		$ Change	% Change
	Non-GMO potato	GMO potato		
Labor	378.1	373.4	-4.6	-2.2
Mechanical/draft power	52.2	52.2	0.0	0
Seed	371.2	389.6	14.8	5
Fertilizer	240.8	259.3	18.5	7.7
Pesticides	38.4	7.7	-30.7	-80
Irrigation	23.7	23.5	0.0	0
Total variable cost	1108.4	1105.8	-2.7	-0.24
Yield(kg/ha)	23950	27543	3593	15
Total return ($/ha)	1842.3	2118.7	276.3	14.99
Gross margin ($/ha)	733.9	1013.1	279.2	38

TABLE 4.
Analysis of economic performance of GMO potato and non-GMO potato in Bogra

Cost Item	Cost ($/ha)		$ Change	% Change
	Non-GMO potato	GMO potato		
Labor	460.1	460.0	4.6	2
Mechanical/draft power	33.3	33.3	0.0	0
Seed	221.0	229.9	8.8	4
Fertilizer	240.0	258.5	18.5	7.7
Pesticides	54.9	7.7	-47.2	-86
Irrigation	18.6	23.5	0.0	0
Total variable cost	1027.8	1056.7	-14.5	-1.39
Yield(Kg/ha)	25840	29716	3876	15
Total return ($/ha)	1987.6	2285.8	298.2	15
Gross margin ($/ha)	959.8	1229.2	283.7	30

TABLE 5.
Analysis of economic performance of GMO potato and non-GMO potato in Munshigonj

Cost Item	Cost ($/ha)		$ Change	% Change
	Non-GMO potato	GMO potato		
Labor	542538.1	4.6	1.9	
Mechanical/draft power	48.3	48.3	0.0	0
Seed	682.7	701.2	18.5	2.7
Fertilizer	399.9	418.3	18.5	4.6
Pesticides	58.9	7.7	-51.2	-87
Irrigation	27.0	23.5	0.0	0
Total variable cost	1759.6	1737.1	-22.5	-1.3
Yield(Kg/ha)	29160	33534	4374	15
Total return ($/ha)	2916.4	3353.4	437.0	14.98
Gross margin ($/ha)	1156.8	1616.3	459.5	40

TABLE 6.
Analysis of economic performance of GMO potato and non-GMO potato in three regions

	Cost ($/ha)			
Cost Item	**Non-GMO potato**	**GMO potato**	**$ Change**	**% Change**
Labor	457.1	452.4	4.6	2
Mechanical/draft power	44.6	44.6	0.0	0
Seed	425.0	443.4	18.5	4.3
Fertilizer	291.3	309.8	18.5	6.3
Pesticides	51.2	7.7	-43.5	-85
Irrigation	23.5	23.5	0.0	0
Total variable cost	1292.6	1281.4	11.1	-0.9
Yield(Kg/ha)	26110	30027	3916.5	15
Total return ($/ha)	2201.1	2531.3	330.2	15.0
Gross margin ($ha)	908.5	1249.9	341.4	37.6

A of partial budget analysis of replacing traditional variety by a LBR transgenic one is presented in Table 7. The introduction of LBR potato would reduce pesticide cost by $37.5/ha and pesticides labor cost by $40/ha. The total incremental benefit is expected to be $867.9/ha against an incremental cost of $46.6/ha. Thus yielding a net benefit of $821.4/ha.

TABLE 7.
Partial budget of LBR potato replacing traditional potato

Particular		**Based on 2005 crop budget ($/ha)**
Incremental benefits		
Reduced cost	Irrigation cost	7.80
	Irrigation labor	3.09
Added return		
	Increased revenue	937
Total incremental benefits		947.89
Incremental costs		
Added cost	Seed cost	2.08
	Fertilizer	3.93
	Harvest labor	7.85
Reduced return		-
Total incremental cost		13.86
Net Incremental benefits		934.03

Distribution of benefits and social return

The results of the economic surplus analysis are presented in Table 8. The introduction of transgenic LBR potato in Bangladesh would have a high social return on investment Table 8 shows the estimated social return and its distribution between producer and consumers for the period 2110 to 2119. The total amount of discounted economic surplus from producing LBR transgenic potato over 10 years is projected at $261 million US dollars with a producer surplus of $87 million and a consumer surplus of $174 million. The internal rate of return (IRR) is estimated at 37% and the net present value (NPV) at $231.6 million.

TABLE 8.
Ex-ante analysis of consumer and producer surplus, net benefits, and social rate of return from producing transgenic potato in Bangladesh, 2010-2019

YEAR	Change in Economic surplus (thousand $)	Change in Consumer Surplus (thousand $)	Change in Producer surplus (thousand $)	Net BENEFIT (thousand $)	NPV (thousand $)	IRR (%)
2005	0.00	0.00	0.00	-9720	231,333.6	36.9
2006	0.00	0.00	0.00	-9720		
2007	0.00	0.00	0.00	-9720		
2008	0.00	0.00	0.00	-9720		
2009	0.00	0.00	0.00	-9720		
2010	16958.99	11305.99	5653.00	16959		
2011	34122.52	22748.34	11374.17	34123		
2012	51490.57	34327.05	17163.52	51491		
2013	51490.57	34327.05	17163.52	51491		
2014	51490.57	34327.05	17163.52	51491		
2015	51490.57	34327.05	17163.52	51491		
2016	51490.57	34327.05	17163.52	51491		
2017	51490.57	34327.05	17163.52	51491		
2018	51490.57	34327.05	17163.52	51491		
2019	51490.57	34327.05	17163.52	51491		
Total	463006.07	308670.72	154335.36	463006		
Discounted Total (at 5%)	260974	173982.75	86991			

Sensitivity analysis

Sensitivity analysis was conducted under alternative scenarios of (i) a 30 percent increase in yield, (ii) a 20 percent increase in input cost, (iii) a 50 percent change in supply elasticity, and (iv) a 50 percent increase in base yield, cost, and supply elasticity. Under the first scenario of a 30 percent increase in yield, compared to the base situation, the IRR and NPV increased from 37% to 54% and $231 million to $516 million, respectively.

In case of a 20 percent increase in input costs, the transgenic technology would still remain viable with a 15% IRR and a NPV of $40 million. In case of 50 percent increase in the supply elasticity, the IRR decreased from 37% to 27% and NPV from $231 million dollar to $124 million. In the case of a 50 percent increase in base yield, input cost and supply elasticity, the IRR, NPV, net benefit, total economic surplus each changed slightly. Thus the transgenic LBR potato technology would remain viable over the exiting non-transgenic one.

Environmental Impact

From our field surveys of farmers, scientists and industry experts, and also from farmers' focus group discussions, it is clear that the growers are aware about late blight disease and are most of them are risk averse, often trying to avoid risk of crop loss by spraying high doses in advance of late blight infection. The frequency of spraying is often 2 to 3 times per week. On the basis of our survey results, it was estimated that total pesticide use in potato production in Bangladesh is 956,670 liters and 1,471,800 kg solids which can be greatly reduced (80%) by using transgenic LBR potato. Thus, the extent of environmental pollution is much higher in the case of non-transgenic potato. Pesticide use is causing pollution of waterways and groundwater, harm to non-target organism and species biodiversity. Pesticide applications are causing a health hazard for farmers, killing beneficial insects, and destroying flora and fauna. Moreover, it has a residual effect in the food chain. Therefore, a safer transgenic technology is desirable for protection of health, environment, and biodiversity. Because LBR transgenic technology results in the use of fewer, less toxic, and persistent insecticides, it should also lead to decrease in negative impacts on human health and non-target organisms.

Conclusion

Potato is the third most important crop in Bangladesh during the winter season. Late blight is considered to be the most important problem of potato in Bangladesh. Farmers use large amounts of chemical pesticides to control the disease. The total quantity of pesticides used amounted to $56 million. There is scope for pesticide cost reduction through adopting a late blight resistant transgenic potato variety.

The minimum expected yield increase of GM potato over Non-GM potato is 15 percent. With this yield change, the gross margin of GM potato in Rangpur would increase by 38 percent ($279/ha) and by 30 percent ($284/ha) in Bogra and 40 percent in Munshiganj ($459.5/ha). However, in all regions the extent of increase in the gross margin for LBR potato over conventional non-GMO potato is expected to be about 37 percent. The commercialization of a specific transgenic LBR potato variety could potentially offer a significant increase in gross margin and profitability to the producers.

Partial budget analysis of replacing traditional varieties by LBR transgenic potato showed that introduction of LBR potato would reduce pesticide cost by $37.5/ha and pesticide labor cost by $40/ha. The total incremental benefit was found to be $867.9/ha against an incremental cost of $46.6/ha, thus yielding a net benefit of $821.4/ha.

The introduction of transgenic LBR potato in Bangladesh would have sizable social benefits. The total amount of discounted economic surplus from producing LBR transgenic potato over a 10 year period (2110 to 2119) would be $261 million, with a producer surplus of $87 million and a consumer surplus of $174 million. The rate of social return was also found to be high, as indicated by an internal rate of return (IRR) of 37 percent and a NPV of $231 million.

The results of sensitivity analysis imply that transgenic LBR potato technology would remain viable over the existing non-transgenic variety under various conditions.It was estimated that total pesticides used in potato production in Bangladesh total 956,670 liter liquid and 1,471,800 kg solids, which can be reduced 80 percent by using transgenic LBR potato. Thus environmental benefits should be large.

Reference

Alston, J.M. , G.W. Norton, and P.G. Pardey (1995). Science under scarcity: Principles and practice for agricultural research evaluation and priority setting, Cornell University Press, Ithaca.

BBS (1985-86, 1992, 1995, 2000, 2004 and 2005). Statistical Year Book of Bangladesh (Various issues), Bangladesh Bureau of Statistics, Ministry of Planning, Government of People's Republic of Bangladesh, Dhaka.

Baset, M.A. (2003). Returns to investment in potato research in Bangladesh, Ph.D Thesis, Department of Agricultural Economics, Bangladesh Agricultural University, Mymensingh, July, 2003.

Chapter **9**

Pod Borer Resistant Chickpea in Bangladesh

S.M. Fakhrul Islam and G.W. Norton

Introduction

Chickpea is the third most important pulse crop of Bangladesh with respect to area (16,298 ha) and production (11,855 tonnes) and contributes about 20% of the total pulses (BBS, 2005) (Table1). It is one of the best legumes for human and animal consumption. Among grain legumes, chickpea is unique because of the variety of food products that are prepared from it in different parts of the world. Although most of the world's chickpea production and consumption is in India (> 70%), the crop is also important in other countries of Asia, Africa, Europe, and America (ICRISAT, 1986).

Chickpea is a temperate crop although it is well adapted to tropical and sub-tropical conditions (Kay, 1979). In the tropics and sub-tropics, it is normally sown in the post monsoon period, during winter season. In Bangladesh, chickpea is grown on well drained alluvial to clay loam soils having a pH 6.0 to 7.0. It is drought tolerant and can be grown in highlands. The acreage of chickpea cultivation in Bangladesh has been decreasing due to a lower return compared to other crops.

TABLE. 1.
Area, yield and production of chickpea in Bangladesh 1990-91 to 2003-2004

Year	Area (ha)	Yield (kg/ha)	Production (tonnes)
1990-91	96757	734	71005
1991-92	92188	702	64680
1992-93	90510	744	67375
1993-94	85626	719	61535
1994-95	84947	730	62035
1995-96	84933	724	61480
1996-97	84435	728	61485
1997-98	84059	713	59900
1998-99	16650	735	12245
1999-00	16445	728	11980
2000-01	16298	727	11855
2001-02	15366	724	11120
2002-03	15144	733	11100
2003-04	13915	746	10380

Sources: BBS 1985, 1986, 1992, 1995 and 2000, 2004

Pod borer is considered the most important problem of chickpea cultivation in Bangladesh. Farmers apply large amounts of chemical pesticides to control this pest, increasing the cost of cultivation and creating potential health and environmental problems. Each year, crop losses due to this pest are around is 30 to 60 percent.

Economic impact information on genetically modified (GM) crops is available for the principal GM crop-growing countries (Commission of the European Communities, 2000; Dowley, et al., 2001; Falck-Zepeda, J.B., Traxler, G., and R.G. Nelson, 2000; Dowley, et al., 2001; Falck-Zepeda, J.B., Traxler, G., and R.G. Nelson, 2000; Tolstrup, et al., 2003;Van Meijl and Tongeren, 2003; Demont, et al., 2004a and 2004b; Alston, J. M, Hyde, J., Marra, M.C., & Mitchell, P.D, 2002). For Bangladesh, such information is not yet available. The purpose of the present study is to provide an assessment of the potential economic impacts of transgenic pod borer resistant (PBR) chickpea in Bangladesh. The objective is to predict costs and/or benefits that a producer might experience if a transgenic PBR chickpea is to be cultivated in Bangladesh, as well as the aggregate benefits to society.

For producers, an economic incentive is an essential factor in decisions to adopt or reject a new technology (eg. GM crops). For this research, it is assumed that producers will base their adoption decision on the relative prices of conventional and GM seeds, chemical pesticides, labour, capital, and other relevant inputs and choose a system that will minimize these costs per tonnes of production. If producers are to be motivated to adopt a GM technology, production costs per tonnes will have to decrease (Kalaitzandonakes, 2003). In addition, farmer diversity, in terms of management ability, agronomic factors, and/or geographic location (Fulton and Keyowski, 1999), will determine the extent of economic gains from GM crops, as producers will differ in their tolerance for risk (Kalaitzandonakes, 2003). Consequently, it can be concluded that some producers will benefit from the new technology while others may not (Fulton and Keyowski, 1999).

Partially offsetting the anticipated cost savings associated with the adoption of a GM technology, the producer will also bear some additional costs. A technology cost typically will be passed to farmers in the form of a seed premium (PG Economics, 2003) as demonstrated in the United States and Argentina (US General Accounting Office, 2000), Spain (Demont and Tollens, 2004b), and the United Kingdom (May, 2003). Some other associated costs such as harvest labor may also increase (Tolstrup et al., 2003).

METHODS

Data

Given the intensity of chickpea cultivation, Rajshahi district was selected for field surveys for the present study. Data were collected from 130 farmers of which 100 were surveyed in the 2004 production season and 30 were surveyed in 2005. Data were also collected from 5 scientists and 8 industry experts from the Department of Agricultural Extension (DAE), the Bangladesh Agricultural Development Corporation (BADC), and seed companies.

Data collected from farmers included various input costs such as seed, fertilizers, pesticides and labor. Data were collected on crop varieties, sources of seeds, crop management practices, input and output prices, crop losses due to pod borer, and crop yields.

From scientists, data were collected on potential regions for GM chickpea, expected yield changes if GM chickpea were grown, changes in variable costs for GM chickpea, time lags and costs for technology development and regulatory requirements.

From industry experts, data were collected on preferred varieties, sources of seed, extent of crop losses due to pod borer, expected extent and time path of adoption of GMO chickpea, time lags and costs for technology development and regulatory requirements.

Economic surplus analysis

Cost and return information, secondary data, and adoption information were combined in an economic surplus model to project total economic benefits and their distribution by region within the country and to consumers, producers, and seed sector (Alston, Norton, and Pardey 1995). The costs of the research and product development (including meeting regulatory hurdles) were included along with the benefits in a benefit cost analysis of the public investment. Excel spreadsheets were used for estimating the economic surplus models.

RESULTS AND DISCUSSION

Trends in chickpea acreage and yield

There was a negative trend in chickpea acreage and production in the selected district of Rajshahi from 1990-91 to 2003-04. Total chickpea acreage of Rajshahi district decreased from 7877 ha to 1949 ha with an annual average decreasing rate of 1.77 percent. Also there was a negative trend in chickpea acreage and production in overall Bangladesh during the same period.

Chickpea production decreased in all the districts as a result of reduction of area during 1990-91 to 2003-04. However, chickpea yield was almost stable around the mean yield of 728 kg/ha during the same period as a result of introduction of new improved varieties of BARI.

Variety cultivated and sources of seed

Sixty percent of the chickpea area in Rajshahi is covered by the HYV variety named BARI5, 13 percent by Nabib, and 27 percent by local varieties. The popular variety of Rajshahi is BARI5 because of its higher yield. On average, 13 percent of chickpea seeds came from farmers' own seeds and 49 percent came from the market in production year of 2005 (Figure 1). However, other important sources of seeds were BADC and NGO.

Insecticides used in chickpea production

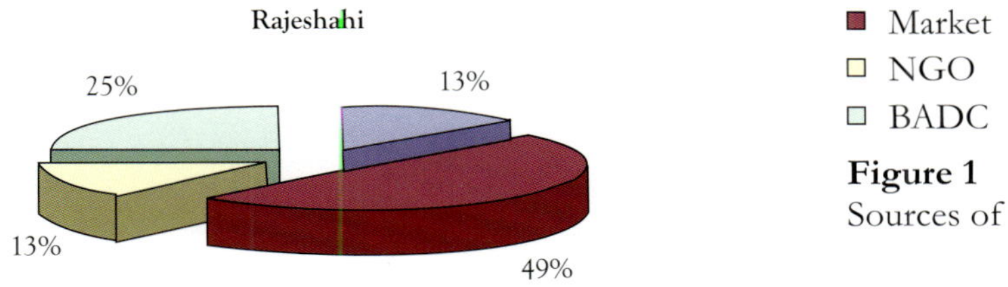

Figure 1
Sources of Chickpea Seed

Almost 100 percent of total insecticides used in chickpea production were for pod borer control in Rajshahi. On average, chickpea farmers in Rajshahi used 310 ml of liquid insecticides per ha for chickpea production. Considering these parameters, the total quantity of insecticides applied in Rajshahi for chickpea cultivation was estimated at 604 liters. Such a quantity of insecticides resulted in an expenditure of 9260 US dollars in Rajshahi district, indicating scope for cost savings in terms of insecticides for chickpea cultivation and for protecting the environment by means of adopting a PBR transgenic chickpea crop.

Cost structure in chickpea production

The largest cost components of chickpea production in Rajshahi were draft power (29% of total variable costs) followed by labour (27%) and seed (25%). The share of insecticide cost was 5 percent of total variable cost of chickpea production.

Economic surplus results

Spreadsheets were used to estimate consumer surplus, producer surplus, total net benefits and rate of return. Research expenditures were assumed to begin in 2005 and to continue until 2009. Thus there is a five years research and technology development lag before benefits occur. The main assumptions for the economic surplus and rate of return analysis were based on field surveys of farmers, scientists, and industry experts. The summary of assumed changes in the chickpea budget due to PBR chickpea are presented in Table 2.

The yield of transgenic PBR chickpea is expected to increase by 15-30 percent. Pesticide cost is assumed to decrease by 100 percent, with slight increase in seed, fertilizer, and harvest labor costs. The probability of success is assumed to be 70-90 percent.

TABLE 2.
Summary of budget changes due to adoption of PBR chickpea

Item	Change
Variable cost change	
Seed cost	decrease by 17%
Fertilizer cost	increase by 4-10%
Irrigation cost	no change
Pesticide cost	decrease by 90-100%
Labor cost	increase by 2-6
Pesticide labor cost	decrease by 90%
Harvest labour cost	increase by 10%
Other labor cost	no change
Other costs	no change
Yield & return change	
Yield (kg)	increase by 15-30%
Price per kg	no change
Gross return	increase by 37-64%
Current yield loss	15-40%
Variety	BARI5, Nabbi
Probability of success	70-90%

Source: BADC, Private

TABLE 3.
Assumptions used in economic surplus model

Parameter	Description and Value
Year	Annual benefits are projected for 15 years after research commences, 2005-2019 (t = 1,2,......,15)
Supply elasticity	The supply elasticity, ε, is set at 0.50 for chickpea.
Demand elasticity	The demand elasticity, η, is set at 0.20 for chickpea.
Proportionate yield change	The estimated yield increase, E(Y), is 50%. Proportionate
change in input cost per hectare	The proportionate change in input cost per hectare is 4%.
Probability of research success	The technology has not been released yet, and the relevant probability of research success is set at 0.80.
Adoption rate	For the purposes of simulation the assigned maximum adoption rates are 30%, 60%, and 90%. Time required to reach the peak: 3 years
Price	Wholesale prices for the period of 2000-2005 are averaged, giving a mean value of $ 415.85 per tonnes.
Quantity	The pre-research quantity is constant, equal to the base quantity 323 thousand tonnes in 2005.
Research cost	The estimated annual research cost for chickpea is 15000 dollar per year.

Analysis of costs and profitability of GM and non-GM chickpea

According to our farmer surveys of 2004 and 2005, pod borer continues to be a major problem in Rajshahi causing annual losses in yield (15 to 40 percent). Present crop regimes require a regular, high rate of insecticide application at short intervals throughout the growing season. If a PBR chickpea is cultivated in Rajshahi, it would reduce insecticide cost per ha by $3.83 to $4.75 per ha according to the survey of 2004, and 2005. The commercialization of SFBR chickpea would reduce labor cost in the region by 6.6 percent. The total variable cost saving in Rajshahi for PBR chickpea would be 3.27 percent according to the survey of 2004 and based on the 2005 survey, it would be 4.52 percent. The average of these two periods would be 3.55 percent (Tables 4, 5 and 6).

The yield of PBR chickpea is expected to increase by 73 kg/ha in Rajshahi according to the 2004 survey, by 152 kg/ha according to the 2005 survey, and by 112 kg/ha averaging both surveys. The minimum expected yield increase of GM chickpea over Non-GM chickpea is 15 percent. With this yield increase, the gross margin of PBR chickpea in Rajshahi would increase by 71 percent ($49/ha) based on the 2004 survey, by 22 percent ($67/ha) based on the 2005 survey and by 30 percent ($58/ha) based on the average of both surveys. The commercialization of a specific transgenic PBR chickpea variety could potentially offer a significant increase in gross margin and profitability to the producers.

TABLE 4.
Analysis of economic performance of GMO chickpea and non-GMO chickpea in Rajshahi, 2004

	Cost ($/ha)			
Cost Item	**Non-GMO potato**	**GMO potato**	**$ Change**	**% Change**
Labor	41.22	38.91	2.31	-6.6
Mechanical/draft power	32.31	32.31	0.00	0
Seed	22.34	23.88	1.54	6.9
Fertilizer	17.57	18.34	0.77	4.4
Insecticides	3.83	0.00	-3.83	-100.0
Irrigation	0.00	0.00	0.00	0
Total variable cost	117.26	113.43	3.83	3.27
Yield(kg/ha)	488	561	73.2	15
Total return ($/ha)	186.98	232.42	45.44	24.0
Gross margin ($/ha)	69.72	118.99	49.27	71.0

TABLE 5.
Analysis of economic performance of GMO chickpea and non- GMO chickpea in Rajshahi, 2005

	Cost ($/ha)			
Cost Item	**Non-GMO potato**	**GMO potato**	**$ Change**	**% Change**
Labor	28.67	26.37	2.31	9.0
Mechanical/draft power	30.88	30.88	0.00	0
Seed	26.00	27.54	1.54	5.9
Fertilizer	14.74	15.51	0.77	5.2
Pesticides	4.75	0.00	-4.75	-100.0
Irrigation	0.00	0.00	0.00	0
Total variable cost	105.05	100.30	-4.75	-4.52
Yield(Kg/ha)	1011	1162	152	15.0
Total return ($/ha)	418.31	481.25	62.95	15.0.
Gross margin ($/ha)	313.26	380.96	67.70	22.0

TABLE 6.
Analysis of economic performance of GMO chickpea and non-GMO chickpea in Rajshahi, 2004 and 2005.

	Cost ($/ha)			
Cost Item	**Non-GMO potato**	**GMO potato**	**$ Change**	**% Change**
Labor	34.95	32.99	2.31	7.6
Mechanical/draft power	31.59	31.59	0.00	0
Seed	24.17	25.71	1.54	6.4
Fertilizer	16.16	16.93	0.77	4.8
Pesticides	4.29	0.00	-4.29	-100.0
Irrigation	0.00	0.00	0.00	0
Total variable cost	111.15	107.21	-3.94	-3.55
Yield(Kg/ha)	749.23	861.61	112.38	0.15
Total return ($/ha)	302.65	356.84	54.19	17.91
Gross margin ($ha)	191.49	249.63	58.13	30

A partial budget analysis of replacing traditional variety by a PBR transgenic one is presented in Table 7. The introduction of PBR chickpea would reduce insecticide cost by $4.29/ha and pesticide labor cost by $1.97/ha. The total incremental benefit is expected to be $60.46/ha against an incremental cost of $3.19/ha, thus yielding a net benefit of $57.27/ha.

TABLE 7.
Partial budget of PBR chickpea replacing traditional varieties

Particular		**Based on 2005 crop budget ($/ha)**
Incremental benefits		
Reduced cost		
	Pesticides cost	4.29
	Pesticides labor	1.97
Added return		
	Increased revenue	54.19
Total incremental benefits		**60.46**
Incremental costs		
Added cost		
	Seed cost	1.54
	Harvest labor	1.65
Reduced return		–
Total incremental cost		13.19
Net Incremental benefits		57.27
Inc B:C ratio		19

Distribution of benefits and social return

The results of economic surplus analysis model are presented in Table 8. The introduction of transgenic PBR chickpea in Bangladesh would have a high social return on investment. Table 8 shows the estimated social return and its distribution between producers and consumers for the period 2110 to 2119. The total amount of discounted economic surplus from producing PBR transgenic chickpea over 10 years is projected at $113 million US dollars with a producer surplus of $32 million and a consumer surplus of $80 million. The internal rate of return (IRR) is estimated at 15% and the net present value (NPV) at $56 million.

TABLE 8.
Consumer and producer surplus, net benefits, and social rate of returns from producing transgenic PBR chickpea in Bangladesh, 2010-2019

YEAR	Change in Economic surplus (thousand $)	Change in Consumer Surplus (thousand $)	Change in Producer surplus (thousand $)	Net BENEFIT (thousand $)	NPV (thousand $)	IRR (%)
2005	0.00	0.00	0.00	-13000.00	56,361.22	15.33
2006	0.00	0.00	0.00	-13000.00		
2007	0.00	0.00	0.00	-13000.00		
2008	0.00	0.00	0.00	-13000.00		
2009	0.00	0.00	0.00	-13000.00		
2010	7614.19	5438.71	2175.48	7614.19		
2011	15491.77	11065.55	4426.22	15491.77		
2012	20889.82	14921.30	5968.52	20889.82		
2013	20889.82	14921.30	5968.52	20889.82		
2014	20889.82	14921.30	5968.52	20889.82		
2015	20889.82	14921.30	5968.52	20889.82		
2016	20889.82	14921.30	5968.52	20889.82		
2017	20889.82	14921.30	5968.52	20889.82		
2018	20889.82	14921.30	5968.52	20889.82		
2019	20889.82	14921.30	5968.52	20889.82		
Total discounted at 5%	112644	80460.31	32184			

Sensitivity analysis

Sensitivity analysis was conducted under alternative scenarios of (i) a 20 percent increase in input cost, (ii) a 50 percent change in supply elasticity, (iii) a 50% change in demand elasticity and (iv) a 50 percent increase in base yield, cost, supply elasticity and demand elasticity. In the case of a 20 percent increase in input costs, the transgenic technology would still remain viable with a 13% IRR and a NPV of $38 million. In the case of a 50 percent increase in the supply elasticity, it was observed that compared to the base scenario, the IRR decreased from 15% to 9% and NPV from $56 million dollar to $19 million. In the case of a 50 percent increase in base yield, input cost and supply elasticity, it was observed that IRR, NPV, net benefit, and producer and consumer surplus all changed slightly. Thus the transgenic PBR potato technology would remain viable over the exiting non-transgenic one.

Environmental Impact

From the field surveys of farmers, scientists and industry experts, and also from farmers' focus group discussions, it is clear that the growers are aware of the pod pest problem and often try to avoid risk of crop loss by spraying high doses in advance of PB attack. On the basis of the survey, it was estimated that total insecticide use in chickpea production in Bangladesh is 956,670 liters and 1,471,800 kg solids which can be greatly reduced by using transgenic PBR chickpea. Thus, the extent of environmental pollution is high in the case of non-transgenic chickpea. Insecticide use is contributing to pollution of waterways and groundwater, harm to non-target organisms and species biodiversity. Pesticide applications are causing a health hazard for farmers, killing beneficial insects, and destroying flora and fauna. Moreover, it has a residual effect in the food chain. Therefore, a safer transgenic technology is desirable for protection of health, environment, and biodiversity. Because PBR transgenic technology results to the use of fewer, less toxic, and persistent insecticides, it should also lead reduce negative impacts on human health and non-target organisms.

Conclusion

Chickpea is the third most important pulse crop grown in Bangladesh during the winter season. Pod borer is considered to be the most important problem of chickpea in Bangladesh. Farmers use large amounts of chemical pesticides to control the problem.

The study revealed that almost 100 percent of total insecticides used in chickpea production were for pod borer control in Rajshahi. On average, chickpea farmers in Rajshahi used 310 ml of liquid insecticides per ha for chickpea production, representing an expenditure of $9260. There is a good scope for pesticide cost reduction by means of adopting a PBR transgenic chickpea crop.

The minimum expected yield increase of GM chickpea over Non-GM chickpea is 15 percent. With this yield increase, the gross margin of PBR chickpea in Rajshahi would increase by 71 percent ($49/ha) based on the 2004 survey, by 22 percent ($67/ha) based on the 2005 survey and by 30 percent ($58/ha) based on the average of both surveys. The commercialization of a specific transgenic PBR chickpea variety could potentially offer a significant increase in gross margin and profitability to producers.

Partial budget analysis of replacing traditional varieties by a PBR transgenic one found that the introduction of PBR chickpea would reduce insecticide cost by $4.29/ha and pesticides labor cost by $1.97/ha. The total incremental benefit is expected to be $60.46/ha against an incremental cost of $3.19/ha, thus yielding a net benefit of $57.27/ha.

The introduction of transgenic PBR chickpea in Bangladesh would have sizable social benefits. The total amount of discounted economic surplus from producing PBR transgenic chickpea over a 10 year period (2110 to 2119) would be $113 million, with a producer surplus of $32 million and a consumer surplus of $80 million. The rate of social return was also found to be high, as indicated by an internal rate of return (IRR) of 15 percent and a NPV of $56 million.

The results of sensitivity analysis imply that the transgenic PBR chickpea technology would remain economically viable over the existing non-transgenic variety under such conditions. Environmental benefits should also be large.

Reference

Alston, J.M. , G.W. Norton, and P.G. Pardey (1995). Science under scarcity: Principles and practice for agricultural research evaluation and priority setting, Cornell University Press, Ithaca.

Alston, J. M, Hyde, J., Marra, M.C., & Mitchell, P.D. (2002). An ex ante analysis of the benefits from the adoption of corn rootworm resistant transgenic corn technology. AgBioForum, 5(3), 71-84. Available on the World Wide Web: http://www.agbioforum.org.

BBS (1985, 1986, 1992, 1995, 2000, 2004 and 2005). Statistical Year Book of Bangladesh (Various issues), Bangladesh Bureau of Statistics, Ministry of Planning, Government of People's Republic of Bangladesh, Dhaka.

Commission of the European Communities (2000). *"Economic impacts of genetically modified crops on the agri-food sector.* Brussels: CEC Directorate-General for Agriculture. Available on the World Wide Web: http://europa.eu.int/comm/agriculture/publi/gmo/fullrep/ackn.htm

Demont, M., & Tollens, E. (2004a). First impact of biotechnology in the EU: Bt maize adoption in Spain. *Annals of Applied Biology,* 145, 197-207.

Demont, M., Wessler, J. &,Tollens, E. (2004b). Biodiversity versus transgenic sugar beet: the one euro question. *European Review of Agricultural Economics,* 31, 1-18.

Dowley, L.J., Leonard, R., Rice, B., & Ward, S. (2001). *Low input fungicide programmes for the control of late blight in potatoes.* Oak Park, Carlow, Ireland: Teagasc Crops Research Centre. Available on the World Wide Web: http://www.teagasc.ie/research/reports/crops/4312/eopr-4312.htm.

Falck-Zepeda, J.B., Traxler, G., and R.G. Nelson, 2000. Surplus distribution from the introduction of a biotechnology innovation. American Journal of Agricultural Economics 82, 360-369.

Fulton, M., & Keyowski, L. (1999). The producer benefits of herbicide-resistant canola. *AgBioForum*, 2(2), 85-93. Available on the World Wide Web: http://www.agbioforum.org.

International Crop Research Institute for the Semi Arid Tropics (ICRISAT, 1986). Annual Report 1986, Patancherru, India.

Kalaitzandonakes, N.G. (2003). The economic and environmental impacts of agbiotech: An introduction. In N.G. Kalaitzandonakes (Ed.), *The Economic and environmental impacts of agbiotech* (pp.1-18). New York: Kluwer Academic.

Kay, D.E. (1979). Food legumes: Crop and product digest, No. 3, Tropical Product Institute, London.

May, M.J. (2003). Economic consequences for UK farmers of growing GM herbicide tolerant sugar beet. *Annals of Applied Biology,* 142, 41-48.

PG Economics. (2003). *Consultancy support for the analysis of the impact of GM crops on UK farm profitability.* Dorchester, Dorset, UK: PG Economics Ltd. Available on the World Wide Web: http://www.teagasc.ie/research/reports/crops/4312/eopr-4312.htm.

Tolstrup, K., Andersen, S.B., Boelt, B., Buus, M., Gylling, M., Holm, P.B., et al., (2003). *Report from the Danish Working Group on the co-existence of genetically modified crops with conventional and organic crops.* Tjele, Denmark: Danish Institute of Agricultural Sciences. Available on the World Wide Web: http://www.teagasc.ie/research/reports/crops/4312/eopr-4312.htm.

United States General Accounting Office. (2000). *Biotechnology: Information on prices of genetically modified seeds in the United States and Argentina* (GAO/RCED/NSIAD-00-55). Washington, DC: US General Accounting Office. Available on the World Wide Web: http://www.teagasc.ie/research/reports/crops/4312/eopr-4312.htm.

Van Meijl, H., & Van Tongeren, F. (2003, August). *International diffusion of gains from biotechnology and the European Union's Common Agricultural Policy.* Paper presented at the 25th International Conference of Agricultural Economists, Durban, South Africa. Available on the World Wide Web: http://www.teagasc.ie/research/reports/crops/4312/eopr-4312.htm.

Chapter **10**

Summary and Conclusions

C. Ramasamy, K.N. Selvaraj and G.W. Norton

Research and development activities have been undertaken for the past four years with the support of the USAID-funded Agricultural Biotechnology Support ProjectII (ABSPII) for the purpose of commercializing products that solve major pest and other problems in India and Bangladesh. The economic impacts of transgenic insect resistant (Bt) eggplant, drought and salt-tolerant (DST) rice, tobacco streak virus resistant groundnut and sunflower in India, and insect resistant (Bt) eggplant, late blight resistant (LBR) potato, pod borer resistant chickpea, and DST rice in Bangladesh are projected to be significant. A summary of the key results found in the various studies is presented in Table1. Costs and benefits are projected over 15 years and discounted at 5% to obtain a net present value for each transgenic product and country.

Not surprisingly (given the amount of production) the largest projected benefits are for drought-and salt-tolerant rice. The sum of discounted benefits, under the most-likely supply elasticity, equals more than $3 billion. Perhaps surprising are the large benefits for TSVR groundnut at more than $1 billion and LBR potato at more than $600 million. All of the transgenic products are projected to earn high returns, with pod borer resistant chickpeas in Bangladesh generating the lowest returns.

A look at some of the key assumptions in Table 2 for each product reveals why certain products rank higher than others. The primary reasons are differences in projected adoption rates, base level quantities, prices, and supply elasticities. The projected yield, cost, and probabilities of success do not differ greatly across products. However, the adoption rate assumed for Bt eggplant in India is 10 percent while it is 50 percent or more for every other product and 70 percent for eggplant in Bangladesh. These rates were based on expert opinions, but some of the assumed adoption rates may be on the high side. The base quantity for groundnuts produced in India is higher than for eggplant and

TABLE .1
Projected Impacts of Transgenic Crops in India and Bangladesh over 15 years

Product	Country	Initial Year of Benefits	Net Present Value (million U.S. dollars)	Book Chapter
DST Rice	India	2012	\$3258 -- \$3343	2
Bt Eggplant	India	2008	\$25 - \$142	5
TSVR Groundnut	India	2012	\$673 -\$1047	3
TSVR Sunflower	India	2012	\$34 -\$91	3
LBR Potato	India	2012	\$613- \$987	4
DST Rice	Bangladesh	2011	\$119 - \$216	6
Bt Eggplant	Bangladesh	2010	\$28 - \$65	7
LBR Potato	Bangladesh	2010	\$124 - \$231	8
PBR Chickpea	Bangladesh	2010	\$19 - \$56	9

TABLE 2.
Key Parameters for Most Likely Scenario

Product	Country	Supply Elast.	Yield Incr. (%)	Cost Change (%)	Prob. of Success (%)	Max Adopt (%)	Disc. Rate (%)	Q. 000 tons	P. \$/ton
DST Rice	India	0.4	25	2	0.68	50	5	28200	295
Bt Eggplant	India	1.0	45	-15	1.00	10	5	8145	145
TSVR Groundnut	India	1.0	20	1	0.77	60	5	8330	836
TSVR Sunflower	India	1.0	20	-3	0.50	53	5	1090	775
LBR Potato	India	0.4	25	6	90	60	5	23160	147
DST Rice	Bangladesh	0.13	20	14	90	70	5	323	184
Bt Eggplant	Bangladesh	0.5	30	0.36	80	70	5	536	123
LBR Potato	Bangladesh	0.60	15	-.32	80	90	5	3296	84.3
PBR Chickpea	Bangladesh	0.50	50	4	80	80	5	59	515

the price is many times higher, which together with high adoption rate accounts for the much higher benefits for groundnuts than Bt eggplant.

The highest returns in Bangladesh are projected for drought- and salt-tolerant rice and late bight resistant potato. In both cases, the high returns are driven by the large base quantities produced. The scientists and regulators in Bangladesh are also surprisingly optimistic about the short length of time is will take (3-4 years from now) to have products ready for commercialization.

Appendix 1

Example of questionnaires used to gather data to assess economic and environmental benefits of developing and commercializing transgenic crops in India and Bangladesh

Farmer Questionnaire -- Bangladesh

Commodity: Salinity and draught tolerant (SDT) rice

Respondent **Interviewer**

Name: ______________________ Name: ______________________

Location: ______________________ Date: ______________________

Type of land tenure:

Education: ______ Age ______
Years in farming: _____
Years in salinity rice cultivation _____

1.What are your current crop management practices for the following:

a) Land preparation?

Land area _______decimal, variety cultivated _______
Number of plough_________ by Bullock/power tiller, Labour used _____(man days) wage ____Tk/md, Power tiller cost ______

b) Crop establishment?
A. Planting:
Seed rate ________Kg/bigha Price of seed ______Tk/kg,
Source of seed:
Purchased from_____, or Own _____
Planting time _________, Labour used Hired: Male ___md, Female ____md,
Own: Male _____md, Female __ md, wage_____ Tk/md

B. Application of manure:
Quantity __________ton/bigha,

Labour used: Own ____md, Hired ____ md

C. Application of fertilizer: Kg/bigha

Quantity: Urea ____, TSP ____, MP ___ , Zinc sulphate ____ Other (specify) ___

Price: (Tk/kg) Urea ___, TSP ____, MP ___ , Zinc sulphate ____ Other(specify)__

D. Mulching: No. ___, Labour used: Own ____md, Hired ___ md, Wage ___Tk/md

E. Weeding: No. ___, Labour used: Own ____md, Hired ____ md, Wage ___Tk/md

c) Water management? Irrigation No. ____, Source ____
Labour used: Own ____md, Hired ____ md, Wage ___Tk/md, Fuel/Rent cost of pump ___Tk/bigha

d) Pest management information:
Name of pest/disease:____

Control method	**frequency** (per cropping season)	**Quantity/bigha**	**Price**	**Brand name**
Chemicals				
Insecticides				
Fungicides				
Biological				
Botanical				
Others (specify)				

Labour used: Own ____md, Hired ____ md, Wage ___Tk/md

e) Harvesting: Time: _____, Labour used: Own: Male ____md, Female: ____md, Hired: Male ____ md, Female ____md, Wage ___Tk/md

2. What was your yield per bigha for this crop last year? Ton/bigha________
and your average over the last 5 years? Ton/bigha___, Output Price Tk/kg _________

3. What was your average annual crop loss (%) due to the target pest

last year? __________ and over the last 5 years? __________________

4. What are the preferred varieties in your area? __________________

5. What is/are your source/s of seeds/planting materials? __________

6. What are your market outlets? (√)
   a. traders ___________
   b. direct selling ___________
   c. contract growing ___________

7. What are the major problems of potato cultivation in your area:
   a. Pest/disease________________________________
   b. Seed problem________________________________
   c. Labour scarcity________________________________
   d. Fertilizer problem________________________________
   e. Irrigation problem________________________________

Industry Expert Questionnaire

Commodity: Salinity and draught tolerant (SDT) rice

Respondent	**Interviewer**
Name: ____________________	Name: ____________________
Occupation: ________________	Date: ____________________

Institutional affiliation (if any): ____________________
Years of experience on the crop: ______

1. The target pest problem of rice that is required to be addressed by biotechnology is salinity and draught problem.

2. What was the average annual crop loss (%) due to the target pest last year?
   ________ and in the last 5 years? __________

3. What are the preferred varieties in your

area?_______________________

4. What are the main sources of seed ?_______________________________

5. How many years will it take to complete the technology development and to meet the various regulatory requirements?

Year	Time Lag (Years)		
	Minimum	Most likely	Maximum
Technology development			
Regulatory:			
Limited field trial			
Multi-location field trial			
Food safety assessment			
Apply for commercialization			

6. What are the chances (%) that the product will pass the regulatory requirements and be commercialized? _____________

7. What is the maximum percentage of crop area expected to be covered by the biotech crop? ___________________How many years will it take to reach that maximum once the crop is commercially released?_________________________

8. What are the expected costs involved in developing the technology and meeting the regulatory requirements?

	Cost (Lac Taka)										
Year	2005	2006	2007	2008	2009	2010	2011	2012	2013	2014	2015
Technology development											
Regulatory											
Limited field trial											
Multi-location field trial											
Food safety assessment											
Apply for commercialization											

9. Do you expect an increase (decrease) in area devoted to the commodity over the next 10 years? If so, by what percent per year ?

Region	Increase or Decrease	Percent Change Per Year

10. What are the constraints of adoption of genetically modified crop in Bangladesh?

f.____

g.____

h.____

i.____

11. What are the probable solutions that you would like to suggest?

a)____

b)____

c)____

d)____

Scientist Questionnaire

Crop: Salinity and Drought Tolerant (SDT) Rice

Respondent	**Interviewer**
Name: ____________________	Name: ____________________
Position: ____________________	Date: ____________________

Institute_______________________________
Specialty: _________________________
Education: _________________________

Years of experience on the crop: ____

1. The target burning problem of the crop is salinity and draught problem

2. a) If a genetically modified crop is developed and adopted to solve above problem (for those farmers who adopt it in the region) what will be the expected yield change ? _______________%

b) What will be most potential regions for the genetically modified crop

____

3. What percent of total variable costs is currently represented by each variable input? What is your estimate of the percent change in cost (per hectare) (if any) for each of the inputs if the genetically modified crop is adopted?

Region:					
		Most Likely Cost Change (Per ha)			
Variable Input	Current Cost Share (%)	Decrease	No Change	Increase	Percent Change
Variable ($/ha)					
Hired labor					
Fertilizer					
Pesticides					
Seeds					
Marketing					
Other					

4. What is the probability (percent chance) of biotech research developing a solution with a commercially acceptable level of effectiveness against the targeted problem?_________________

5. How many years will it take to complete the technology development and to meet the various regulatory requirements?

Year	Time Lag (Years)		
	Minimum	Most likely	Maximum
Technology development			
Regulatory:			
Limited field trial			
Multi-location field trial			
Food safety assessment			
Apply for commercialization			

6. What are the expected costs involved in developing the technology and meeting the regulatory requirements?

Cost (Lac Taka)											
Year	2005	2006	2007	2008	2009	2010	2011	2012	2013	2014	2015
Technology development											
Regulatory											
Limited field trial											
Multi-location field trial											
Food safety assessment											
Apply for commercializati on											

7. Which variety do you intend to put this technology? (check √ your answer(s)

	Options for variety choice		Remarks
Variety type	Hybrid	Saved seeds/OP	
Variety source	Public	Private	
Variety use	Fresh	Processed	
Target market	Domestic	Export	

8. What are the expected unintended environmental effects? (check √ if a concern)

   a. gene flow ______________________________
   b. reduced biodiversity ______________________
   c. harms non-target organisms ________________
   d. others (specify) __________________________

9. What are the constraints of adoption of genetically modified crop in Bangladesh?
   a. ______________________________
   b. ______________________________
   c. ______________________________
   d. ______________________________

10. What are the probable solutions that you would like to suggest ?
    a) ______________________________
    b) ______________________________
    c) ______________________________
    d) ------------------------------